Quantum Advantage™

Be Ahead. Be Safe. Win.

MIKE MIKHAIL

Quantum Advantage LLC, www.qadva.com

ISBN: 979-8-9932003-0-9

DEDICATION

To those who did instill in us, generation after generation, the curiosity and inquisitiveness. To the ones around us that encourage and support, however hard for them to deal with the process and consequences.
My predecessors, my family, friends, colleagues.
Thank you.

CONTENTS

PREFACE

Quantum technology is no longer a curiosity of physics labs. It is accelerating into the world of business, and its impact will be profound. The organizations that prepare now will enjoy resilience, trust, and opportunity. Those that wait will face disruption, costs, and possibly extinction.

This book is not written for researchers or engineers. It is written for leaders: the executives, board members, and strategy owners who must navigate uncertainty and make decisions today that will still matter a decade from now.

You don't need a PhD in quantum mechanics to lead in this space. You need clarity, urgency without hype, and a plan you can execute.

That's what this playbook provides:

- **Be Ahead** — see the wave before others do, and act before the pack.
- **Be Safe** — protect your foundations before they collapse.
- **Win** — turn disruption into competitive advantage.

We are at a moment of convergence. The first **post-quantum cryptography (PQC) standards** are finalized. The first **quantum pilots** are running in finance, healthcare, and telecom. And the first adversaries are already **harvesting encrypted data**, betting they will decrypt it later.

This book will not overwhelm you with theory but will rather show you how to:

- Recognize the risks clearly.
- Mobilize your teams on practical steps.
- Spot the opportunities that quantum will unlock.

The elephants — slow, massive, revenue-driven incumbents — will hesitate. The cheetahs — lean, fast, decisive organizations — will sprint ahead.

This book is your invitation to run with the cheetahs.

DISCLAIMER

This book summarizes information that is publicly available from standards bodies, government agencies, and industry reports. References to organizations such as NIST, NSA, ENISA, NCSC, MAS, and others are included for educational purposes only.

While every effort has been made to ensure accuracy, any errors or interpretations are solely the responsibility of the author. This book is not affiliated with or endorsed by any government agency, standards body, or corporation mentioned.

This book does not provide legal, compliance, or investment advice. Readers are encouraged to consult qualified professionals and official publications when making decisions related to cryptography, cybersecurity, or quantum technologies.

PART I
WHY QUANTUM ADVANTAGE NOW

This section sets the stage. It explains why quantum computing is no longer a distant curiosity, why leaders cannot afford to wait, and how agility separates tomorrow's winners from those left behind.

1
THE COMING WAVE

The **quantum computing revolution** is no longer a distant vision – it's unfolding now. Global research and industry leaders report a steady stream of breakthroughs: new chip prototypes addressing error rates and scalability, and powerful quantum algorithms being tested on real hardware. In fact, the United Nations declared 2025 the *International Year of Quantum Science and Technology*. Governments around the world are pouring billions into quantum R&D, and fresh public-private initiatives are announced monthly. This surge of momentum – from cutting-edge labs to boardrooms – signals that **quantum is on its way to becoming as transformative as the Internet or AI**. We're entering a new era where quantum advantage is moving from theory to reality.

Quantum computers operate on principles unlike any previous technology. Their promise – solving certain problems exponentially faster than today's machines – is too big for any one company or country to miss. The *real risk* is standing still. As with the internet and cloud computing before it, early experiments may not show immediate ROI, but **waiting for certainty is a high-risk strategy**. The history of innovation shows that disruptive technologies often appear overhyped and confusing at first, then suddenly surge and upend industries. Quantum is primed for a similar breakout. Indeed, world leaders describe quantum as a "game-changer" for sectors from pharmaceuticals to energy, and legislation is moving rapidly to keep pace.

Global Signals: Breakthroughs and Policy

- **Major breakthroughs.** In the past year alone, teams at leading tech labs unveiled new quantum processors and algorithms that push the boundaries of what's possible. For example, announcements of processors with ~100+ qubits (with greatly reduced error rates) demonstrate that **hardware is scaling up**. One report notes that "multiple new quantum chip prototypes" targeting core challenges (error rates, qubit coherence, scalability). These advances suggest practical quantum advantage **may emerge in years, not decades**.
- **Policy momentum.** Governments see quantum as an economic and security imperative. In late 2024 the U.S. Senate unanimously backed a reauthorization of the National Quantum Initiative, authorizing $2.7□billion over five years to shift focus toward applied quantum science. Other nations are moving too. For example, the UK announced a £130□million (≈$162□million) investment in 2025 for quantum technology research on cybersecurity, fraud, and other challenges. **National strategies and funding programs** are multiplying across the EU, US, China, and beyond. This flood of investment – both public and private – sends a clear signal that "quantum is here, and big."
- **Early pilots.** Savvy organizations are quietly experimenting today. Financial firms, pharmaceutical companies, and manufacturing giants are partnering with quantum startups or research labs to test use cases. Typical pilots include portfolio or risk optimization in banking, molecular simulations in drug discovery, and complex logistics or scheduling problems in supply chains. For instance, one global bank reported using a quantum annealer to analyze inter-bank risk in seconds performing a task that would take years classically. In industry after industry, exploratory projects are under way. These pilots send a message that **quantum's practical impact is being felt now**.

Rising Risks: The Impending Cryptographic Threat

Meanwhile, a new kind of risk is ripening. **Most of today's encryption – securing data from banking to health records – can be broken by a sufficiently powerful quantum computer**. Experts warn that malicious actors may already be **"harvesting" encrypted**

data today with the intention of decrypting it years from now when quantum hardware catches up. This "harvest-now, decrypt-later" scenario means **long-lived secrets** (think archives of emails or state documents) are vulnerable if not addressed immediately. In fact, NIST estimates a quantum computer capable of cracking standard encryption could appear *"within a decade"*.

Governance is catching up. In mid-2024, NIST finalized a suite of **Post-Quantum Cryptography (PQC)** standards and urged organizations to start migrating to these new algorithms now. These PQC algorithms are designed to resist quantum attacks but implementing them will be a multi-year effort. As one official put it, system administrators are **"encouraged to incorporate [PQC algorithms] into their systems immediately, because full integration will take time"**. Similarly, international cybersecurity agencies (e.g. NIST and ENISA) are outlining directives to phase in quantum-resistant encryption.

Callout: *Encryption is the foundation of digital trust. A quantum break could undermine entire systems – from finance to healthcare. Forward-looking leaders must act on "crypto-agility" now, not later.*

In summary, the **clock is ticking**. Every day that sensitive data remains protected by quantum-vulnerable encryption increases future exposure. Organizations should inventory their critical data, identify where quantum threats may emerge, and plan for early adoption of PQC and quantum-safe communications.

Emerging Opportunities: Quantum Use Cases

Despite its challenges, quantum computing offers **tremendous upside** across industries. Early adopters are already uncovering advantages in modeling, optimization, and security:

- **Finance & Insurance:** Quantum algorithms excel at complex calculations on large datasets. Banks and insurers are piloting quantum for portfolio optimization, fraud detection, and enhanced risk forecasting. For example, one report describes a pilot where a quantum optimizer rapidly evaluated thousands of economic scenarios to flag systemic risks – work that would have been intractable on classical systems. Quantum-generated insights could fundamentally change trading strategies, risk management, and beyond.
- **Pharmaceuticals & Chemicals:** Drug discovery and

material science are inherently quantum problems – molecules obey quantum physics. Big pharma firms are exploring quantum-enhanced simulations to predict molecular behaviors faster and more accurately than today's computers allow. In one survey, roughly half of life-sciences companies said they plan to invest $2–$10 million per year in quantum research over the next five years. Quantum-enhanced generative algorithms can scan vast chemical spaces for new drug candidates or simulate battery chemistry for longer-lasting cells. These capabilities promise to **compress R&D timelines** and unlock innovations (e.g. novel medicines or sustainable fuels) unreachable by classical methods.

- **Manufacturing & Logistics:** Complex scheduling, supply-chain optimization, and advanced materials design are other quantum sweet spots. Aerospace and automotive players, for instance, have formed consortia to tackle intricate design problems. A recent collaboration among a major aircraft manufacturer, an automaker, and a quantum firm used quantum algorithms to model chemical reactions in hydrogen fuel cells, aiming to develop lighter, more efficient materials for next-generation aircraft. Similarly, manufacturers and retailers are piloting quantum tools to optimize global supply networks and production schedules with millions of variables – tasks that would overwhelm conventional computing.
- **Energy & Infrastructure:** Utilities and energy companies are eyeing quantum to optimize grid operations and materials for batteries or fusion. In power markets, quantum models may better predict demand and optimize generation in real time. Quantum sensors (another branch of the technology) can measure gravitational or electromagnetic fields with extreme precision, which could improve navigation, mineral exploration, or even earthquake sensing. Governments also see quantum sensing as a national security asset for submarine detection, satellite navigation, and more.
- **Cybersecurity & Communications:** Beyond breaking encryption, quantum tech can **strengthen** it. Quantum key distribution (QKD) enables theoretically unbreakable encryption by transmitting quantum states as keys. Telecom and government networks are starting pilots of quantum-secure links. At the same time, early movers in **post-**

quantum cryptography are retooling software and hardware to meet coming standards. For example, major banks have begun upgrading their systems with PQC algorithms and even deploying quantum-random-number generators for secure key creation. These efforts not only guard against future threats, but also lay the foundation for new business models (e.g. quantum-secure financial transactions).

Each of these areas is nascent, but the **direction is clear**. As one analysis notes, "quantum computing is widely accepted as a disruptive technology emerging now". The earliest commercial benefits may be small, but the breakthrough potential is enormous. Crucially, the window for **"early-mover advantage"** is open. Organizations that start exploring quantum applications today can build the talent, partnerships, and data pipelines needed to capitalize when the tech matures. Even modest experiments in optimization or simulation can yield insights and brand leadership.

2
STRATEGIC IMPERATIVE: ELEPHANTS AND CHEETAHS

In this quantum era, enterprises face a choice: be the slow-moving **"elephant"** or the nimble **"cheetah."** Large incumbents often move cautiously – waiting for clear benchmarks or for others to take risks. This is understandable; quantum's jargon and rapid hype cycle can be confusing. But staying still invites danger. Just as giant corporations once scoffed at the Internet only to be disrupted, today's market leaders risk being outpaced by smaller, tech-savvy challengers.

- **Cost of delay.** Research warns that the consequences of inaction can far outweigh the costs of early experimentation. The Internet and cloud computing "followed similar patterns: unclear short-term returns, followed by sweeping disruption". Quantum may likewise transform industries before most leaders realize it. Every quarter spent idle is a quarter where competitors are learning, hiring talent, and forging partnerships.
- **Agility as advantage.** Mid-size firms and startups – unburdened by legacy IT or bureaucratic reviews – can pivot quickly into quantum initiatives. They adopt hybrid strategies, combining quantum and classical systems, to get incremental performance gains. For example, a forward-looking CFO might integrate a quantum solver from the cloud into existing analytics to generate higher-quality solution candidates, then use AI and classical compute to refine them. This hybrid approach amplifies decision-making speed and efficiency far beyond what either system could do alone.

- **Leadership mindset.** Being a cheetah doesn't mean reckless spending; it means committed learning. Many experts advise appointing a C-level "quantum sponsor" to shepherd strategy, align pilots with business goals, and work across functions. Such leaders focus on specific use cases – say, a "quantum lab" for supply-chain optimization – rather than chasing abstract hype. They cultivate partnerships with quantum providers, cloud platforms, and academic labs. They also prepare IT and security teams for the quantum transition (upgrading crypto libraries, training developers in PQC, etc.).

Callout: *"Quantum computing is a strategic imperative, not a curiosity." Leaders who treat it as optional may find themselves outmaneuvered. The current window – before quantum is everywhere – is the time to stake a claim.*

Contrast this with elephants: huge firms with slow procurement, siloed strategy, and risk-averse cultures. Such organizations may believe they can "wait and see." But by the time quantum is a proven utility, it may be too late to catch up. Talent will have been absorbed by aggressive adopters, and regulatory standards (for crypto, for example) may force costly retrofits. In cybersecurity alone, failure to act now could expose customer data, intellectual property, or critical infrastructure to future breaches.

Elephants vs Cheetahs

Elephants	Cheetahs
Moves slowly; waits for full certainty	Moves fast; acts on early signals
Bureaucratic, layered approvals	Agile, empowered teams
Short-term metrics (quarterly revenue)	Long-term advantage, learning-driven
Avoids early investment; reactive	Invests early; proactive experimemntation
High risk of being left behind in quantum race	Secures quantum advantage; market leader

Transform your organization's posture

On the Precipice: The Cost of Inaction

The coming quantum wave carries both **reward and risk**. Companies that harness it gain new capabilities and efficiencies; those that ignore it risk falling behind or vulnerable. The payoff for acting early can be huge: securing long-term competitiveness and innovation leadership. The consequences of complacency can be dire – imagine competitors cutting R&D cycles in half, or attackers one day decrypting your long-archived secrets.

This chapter has set the stage: quantum is real, accelerating, and multi-dimensional. We've seen macro signals (from government budgets to chip launches) and sketched the broad risk/opportunity landscape. The metaphor is clear: **don't be the elephant left behind**.

In **Part II**, we will drill down into what's truly at stake if executives remain passive. We'll quantify the risks to data security, reveal how competitor organizations could pull ahead, and offer a roadmap for building quantum readiness. For now, the message is urgent: *the wave is coming. It's time to prepare.*

Key Takeaways:

- Quantum computing is emerging rapidly, with breakthroughs in hardware and algorithms suggesting that practical advantages are near.
- Major governments and industries are investing billions and launching pilot projects. This signals both opportunity and threat.
- Long-term data encrypted today may be exposed in the future; moving to post-quantum cryptography is now a strategic security priority.
- Early use cases span finance (risk/fraud analysis) , healthcare (drug design) , manufacturing (materials, logistics) and security (unbreakable encryption).
- Leadership matters: nimble "cheetah" organizations are beginning to experiment and adapt, while slower "elephants" face increasing peril.
- **Action:** Begin assessing quantum risks and opportunities in your sector today. The next chapters will guide how to translate this urgency into strategy.

PART II
THE QUANTUM THREAT

Here we confront the risk head-on. These chapters show how quantum computing threatens today's digital trust. They also outline the global standards and mandates emerging and highlight the urgent signals every executive must act on.

3
DIGITAL FOUNDATIONS AT RISK

In the digital era, trust is built on a fragile foundation: cryptography. Every secure online transaction, confidential message, or verified digital signature depends on encryption working as intended. Organizations have come to rely on this invisible **"glue of digital trust"** that holds our connected world together. But today, an emerging technology – **quantum computing** – threatens to dissolve that glue, putting our digital foundations at risk. This chapter explains in business terms why encryption matters, how quantum computing endangers the current model of digital trust, and why boards and CISOs must treat this as an urgent strategic risk.

Encryption: The Glue Holding Digital Trust Together

Modern business runs on encrypted networks and verified digital identities. Encryption isn't just a technical nicety; it is the **backbone of digital security and trust**. Three core elements underlie all secure digital interactions: **confidentiality, identity, and integrity**.

- **Confidentiality (Encryption):** Encryption safeguards data by **"cloaking it in a veil of secrecy"**, ensuring that sensitive information can only be read by intended parties. For example, when you see a padlock icon in your web browser, a protocol called **TLS (Transport Layer Security)** is at work, using algorithms like **RSA** or **ECC (Elliptic Curve Cryptography)** to encrypt the communication. This means even if someone intercepts the data, they can't read it. VPNs (virtual private networks) similarly encrypt data flowing between remote workers and corporate networks. In effect, encryption functions as the locks on the doors of our digital house – keeping intruders out while allowing safe access for authorized users.

- **Identity (Authentication):** Encryption alone isn't enough; we also need to know *who* we are communicating with. **Public Key Infrastructure (PKI)** is the system that links cryptography to identity. It issues digital certificates (like electronic ID cards) to websites, companies, or people, and those certificates are backed by cryptographic keys (often RSA or ECC keys). This allows us to verify that a website or message is authentic and from a trusted source. **Digital signatures** (also powered by RSA/ECC math) are the seals of authenticity that ensure, for instance, an email or a software update actually comes from your IT department and not an imposter. Identity verification mechanisms – from passwords and two-factor logins to biometric scans – all ultimately rely on cryptographic protocols to prevent impersonation. In essence, PKI and digital certificates act as **digital passports** for machines and code, assuring that "entities involved in digital interactions are who they claim to be".
- **Integrity:** Finally, encryption and signatures preserve the **integrity** of data, meaning that information cannot be altered undetectably. If a signed document or piece of software is tampered with, the digital signature breaks, alerting us to the corruption. Integrity is what gives confidence that data – whether a financial record or a software update – remains unmodified from its source. In business terms, integrity is the guarantee that what you sent or stored is exactly what you receive when needed, with no surreptitious changes along the way.

Together, **confidentiality, identity, and integrity create a secure framework that underpins the fabric of digital trust**. For executives, this isn't just IT jargon: it's the quiet engine enabling e-commerce, online banking, telemedicine, cloud computing – virtually every digital business model. If encryption is the glue holding this trust framework together, one can imagine the chaos if that glue fails.

Executive Signal: If your **PKI fails**, your **digital identity collapses.**

Every day, countless examples of this framework in action go unnoticed – which is exactly the point. When an online customer accesses their bank account, **TLS/SSL certificates** verify the bank's identity and encrypt the web session, keeping account details private. When a manager signs a contract digitally, cryptographic signatures validate that the document is genuine and untampered. When employees download a software update, **code-signing certificates** ensure the update is legitimate and hasn't been corrupted by malware. In short, cryptography is deeply embedded in **identity, confidentiality, and authentication** across business operations. It's the **invisible guardian of digital trust** – and it works so well that we rarely think about it.

The Quantum Threat: Cracks in the Foundation

Enter **quantum computing** – a revolutionary technology that, while promising immense benefits in fields like drug discovery and optimization, poses an unprecedented threat to cybersecurity. Quantum computers operate on fundamentally different principles of physics, allowing them to solve certain mathematical problems **magnitudes faster** than traditional computers. Unfortunately, some of those mathematical problems are the very ones that keep our encryption schemes secure.

Today's encryption (such as RSA, ECC, and Diffie-Hellman key exchange) relies on "one-way" math problems: extremely hard puzzles that classical computers would take eons to solve (for instance, factoring a large prime number). This is why, as of now, cracking a 2048-bit RSA encrypted message by brute force is practically impossible. **Quantum computers change that**. With algorithms like **Shor's algorithm**, a powerful quantum computer could factor those large numbers exponentially faster than a classical machine. Tasks that would take a classical supercomputer **billions of years** might take a quantum computer **days or hours**, instantly undermining the security of RSA and ECC encryption. In effect, quantum computing can **solve the underlying mathematical puzzles** much faster, **rendering many current encryption methods (like RSA and ECC) obsolete**.

What does this mean in business terms? It means the locks on our digital doors could spring open without warning. Sensitive data once thought safe for decades could become an open book. Encrypted customer records, financial transactions, intellectual property, or confidential board communications – all would be readable if intercepted by someone with a large quantum computer. Likewise, an attacker could **forge digital signatures** with impunity. They could pretend to be your CEO, your bank, or your software vendor by digitally signing malware or fraudulent transactions that look legitimate. **The entire digital trust model of authentication and identity would be upended.** As one industry expert succinctly put it, *"Quantum computing will upend the security infrastructure of the digital economy"*.

What Breaks, When, and Why It Matters Now

It's important to understand **what exactly would break** and **how soon** this could happen. The critical vulnerability lies in **public-key cryptography** – the RSA, ECC, and Diffie-Hellman algorithms that underlie most encryption key exchanges and digital signatures today. **Asymmetric** cryptography (public/private key pairs) is especially at risk of *complete* compromise by quantum attacks. **Symmetric** cryptography (shared secret keys, like AES encryption) is less directly threatened – quantum techniques could weaken it (by effectively halving the strength of encryption, per Grover's algorithm) but not outright break it. In practice, doubling key sizes can counter most symmetric threats. Thus, the **real Achilles' heel is our dependence on RSA/ECC** for securing transactions and identities. Those algorithms protect "**everything from SSL/TLS certificates to digital signatures**" in web commerce, banking, VPNs, encrypted email, and more. When they fall, the **entire pyramid of digital trust built on them comes into question**.

When will this happen? No one can predict the exact timeline for a quantum computer powerful enough to crack RSA/ECC. Estimates range from a few years to a couple of decades. Major firms and government agencies increasingly believe the threat could materialize by the **early 2030s** – in fact, **60%+ of large enterprises expect quantum computers to be mainstream by 2030**. Some experts warn a **cryptographically relevant** quantum machine (one capable of breaking 2048-bit RSA) could arrive within **10-15 years or sooner. In executive terms: the quantum threat is not a vague someday issue; it is likely to hit within strategic planning horizons**. It could even happen faster than anticipated, especially if secret breakthroughs occur. Notably, **76% of U.S. business leaders in one survey said it's "only a matter of time" before cybercriminals use quantum to decrypt today's data**.

However, smart executives will not wait for a specific "Q-Day" countdown. The **threat matters now** because of a risk known as **"Harvest Now, Decrypt Later."** In this scenario, adversaries don't need a quantum computer in hand; they only need patience. Attackers can **harvest encrypted data today – sensitive databases, intercepted network traffic, encrypted backups – and simply hold onto it.** Years down the line, once they obtain access to a quantum decryption capability, they decrypt that trove of data. The victims might not even know their long-ago stolen data will suddenly become readable. **Secrets that seem safe now (because they're encrypted) may be stolen in 2025 and decrypted in 2035**, exposing business-critical information that is *still* sensitive. As NIST warns, *"some secrets remain valuable for many years…capture encrypted data now in hopes a quantum computer will break it later"*. This is a very concrete near-term risk: even before quantum attacks exist, our data is being vacuumed up under the assumption that it will be readable eventually.

Executive Signal: Long-lived data = Long-lived risk.

Any data with a **long shelf life**, such as **health records**, **personal identifiable information (PII)**, **intellectual property designs, legal documents, or state secrets**, is particularly at risk of this harvest-now/decrypt-later threat. If your business deals with information that needs to remain confidential for years, you have a ticking clock on protecting that information. An executive in healthcare or finance, for instance, must assume that **patient records or credit histories stolen today will still be sensitive in a decade** – meaning attackers have a strong incentive to collect it now and decrypt it when technology allows. This dynamic creates a **business liability time bomb**: a data breach that might seem contained (because the data is encrypted) could *detonate years later* when that encryption is rendered useless.

In summary, the quantum threat model shows cracks forming in our digital foundation. It's not just the far-off specter of an encryption apocalypse; it's a **present-day strategic risk.** Leading security agencies are sounding the alarm. As one U.S. government cybersecurity advisory warned, organizations **"should begin now to plan their transition to 'Quantum Readiness'"** as part of security and business continuity strategies. Forward-looking boards and CISOs must treat this as seriously as any emerging market disruption – because it fundamentally is one, aimed at the core security model of the digital economy.

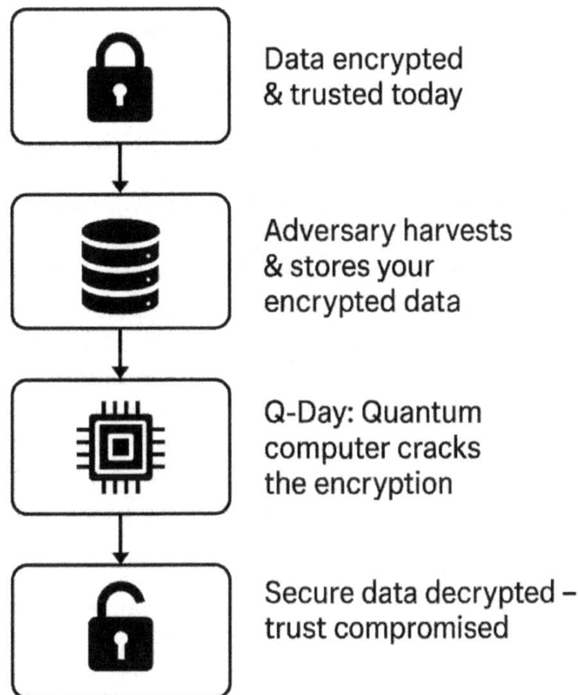

What's Vulnerable? Systems, Data and Trust at Stake

Quantum vulnerability isn't limited to one system or another – it's pervasive. Any application of today's encryption could become a weak link. Here are key areas and assets at risk:

- **Secure Communications (TLS/VPNs):** Protocols like TLS/SSL that secure web browsing, APIs, and mobile app traffic would be **immediately broken** by a quantum adversary. An attacker could **decrypt HTTPS traffic, VPN tunnels, or secure messaging** streams in real time. Imagine competitors reading your company's encrypted cloud transactions or hackers eavesdropping on VPN communications with impunity. The confidentiality of customer data in transit, payment transactions, and proprietary information exchange would vanish. Furthermore, encrypted past communications that are recorded (by governments or hackers) could be retroactively decrypted, exposing archives of supposedly private emails or transactions.

- **Digital Identity & Authentication:** The **public key infrastructure (PKI)** that authenticates users, devices, and websites faces an **"existential threat" from quantum computing.** If an attacker can derive your private key from your public key (which Shor's algorithm enables), they can impersonate you or your systems. **Digital certificates** – the identity cards of the internet – could be forged. This means a hacker could present a fake certificate that computers would accept as genuine, allowing undetected **man-in-the-middle attacks** on secure networks. **Digital signatures** would no longer be trustworthy; contracts, financial transactions, or software updates signed with broken algorithms (like RSA/ECDSA) could be faked. In practical terms, **if your company's certificate authority or signing keys are compromised, attackers can make any malicious message or software appear authentic.** The trust chains that link customers to your website or connect employees to corporate services would be broken. As a result, **digital identity collapses** – your customers, partners, and employees can no longer be sure "you are who you claim online."
    - *Example:* Picture a scenario where a hacker uses a quantum computer to forge the **code-signing signature** on a critical software update for your product. The update, now effectively malware, is delivered to all your customers' devices *with a valid-looking signature.* Their systems install it, believing it's from you, and suddenly your product base is compromised. This is not sci-fi; it's a direct consequence of broken digital signature schemes. The brand damage and liability from such an event could

be devastating.

- **Long-Lived Data Repositories: Stored, encrypted data** is especially vulnerable due to the **"harvest now, decrypt later"** tactic discussed. Any system housing long-term sensitive data is a target. This includes archives of **customer data, credit card numbers, health records, legal files, intellectual property designs, M&A documents**, etc. Even if these databases are stolen in encrypted form, **quantum decryption can later lay them bare. Cloud storage** buckets, backup tapes, or any data at rest that is encrypted with traditional algorithms may be opened in the future. The implications span every sector:
  - In **healthcare**, decades of patient records are meant to remain confidential for a lifetime; a quantum breach could violate privacy on a massive scale and flout regulations like HIPAA.
  - In **finance**, think of long-term loan data, investment histories or private wealth information – a treasure trove for fraud and insider trading if decrypted later.
  - For **tech and manufacturing**, trade secrets or R&D data from years past could suddenly leak to competitors or hostile nations, eroding competitive advantage.
  - **Government and defense** records, some classified for 25+ years, would obviously be at risk, with national security implications.

The business risk here is not just direct exposure, but also **regulatory and contractual liability**. Regulations worldwide (GDPR in Europe, various data protection laws elsewhere) mandate protection of personal data. If encrypted personal data today is compromised tomorrow, regulators may still hold organizations accountable for the breach. It's a compliance nightmare: you thought data was safe under encryption, but quantum computing changes the calculus, potentially leading to fines and legal action years after the initial data theft. In essence, **encrypted data is not "safe" data if your encryption method has an expiration date.**

- **Infrastructure, IoT, and Supply Chain:** Perhaps one of the most overlooked areas is the vulnerability of **embedded**

systems and IoT (Internet of Things) devices. These "smart" devices – from industrial control systems and smart grid sensors to connected appliances and medical devices – often have a long operational life (5, 10, 20+ years) and limited ability to update their cryptography. Many are secured (today) by lightweight cryptographic algorithms that *will* fall to quantum attacks. The **challenge is that these devices can't easily be upgraded or replaced.** If someone can break the encryption or signatures they rely on, they could **inject false data or malicious commands into critical infrastructure** networks. Consider industrial IoT sensors in a power grid or manufacturing line: a quantum-enabled attacker might fake sensor readings or disable security updates by presenting forged credentials. **Firmware updates** are another weak point – many devices accept new software only if it's digitally signed by the vendor. If that signature method (say, ECDSA) is broken, adversaries can load **malicious firmware** onto devices , causing potentially catastrophic failures or safety hazards. The **integrity of the supply chain** – from software code to hardware components – relies on cryptographic trust. Quantum computing could enable subversion of that chain by making it trivial to fake certificates and signatures used in manufacturing and logistics.

It's worth noting the **systemic risk** here: because cryptography is **ubiquitous**, a quantum break undermines security **"through entire technology ecosystems".** The issue is not isolated to one IT system – it is across all systems. Think of it as a single point of failure on a global scale. If we lose confidence in encryption, we lose confidence in **every digital transaction and record.** That is why this has been aptly called a **"digital Y2K moment,"** except this time the threat isn't a date bug but a physics breakthrough. And unlike Y2K, we **don't know exactly when** the critical moment will hit – only that the clock is ticking.

Impact of Cryptographic Failure: Business Risks and Repercussions

For business leaders, the failure of encryption is not just a technical problem, it is a **full-blown business crisis.** Consider the possible impacts if your organization's cryptography were suddenly compromised:

- **Data Breaches and Intellectual Property Loss:** Encrypted databases and communications could be decrypted, leading to massive data breaches. Personal customer data, if exposed, can result in costly **notification efforts, credit monitoring, lawsuits, and fines**. Intellectual property or trade secrets, once lost, can eliminate competitive advantages or derail product strategies. The **Global Risk Institute** warns that quantum computers may crack encryption **faster than anticipated**, implying data breaches on an unprecedented scale if we remain on current tech. The **financial impact** of such breaches could run into the hundreds of millions (from loss of business, legal costs, etc.), not to mention **shareholder lawsuits** if negligence is perceived.

- **Regulatory and Compliance Violations:** Many industries have strict data protection regulations. If sensitive data thought secure becomes exposed, companies could find themselves in violation of laws like GDPR, HIPAA, or sector-specific cybersecurity mandates. Regulators are unlikely to accept "but it was encrypted" as an excuse if standards bodies have long warned of quantum risks. We are already seeing regulators take note – e.g., the U.S. passed a law in 2022 requiring federal agencies to start preparing for post-quantum security. Regulators in financial services and healthcare are issuing guidance that quantum readiness is part of due diligence. Non-compliance could result in **fines, sanctions, or loss of licenses**. In short, quantum-induced breaches could create a compliance nightmare, where suddenly **your security controls are deemed inadequate by evolving standards**.

- **Business Operations Disruption:** Trust is fundamental to operations. If your **digital certificates** (used in everything from authenticating internal systems to customer-facing websites) are no longer trustworthy, you might have to **halt operations** to avoid security catastrophes. Imagine a scenario where all your customers' browsers start warning that your website is insecure (because the certificate could be forged), or your employees can't establish VPN connections because the encryption is broken. The business could grind to a standstill until new secure systems are in place. Additionally, **critical infrastructure** operators (power grids, telecommunications, transportation systems) could face disruptions or sabotage if control systems are compromised. The downtime and safety implications here are severe: this goes beyond data loss to **public safety and service continuity** issues.

- **Reputational Damage and Loss of Trust:** Trust, once broken, is hard to repair. If a quantum attack leads to a high-profile failure, say, customer data from years past suddenly leaks, or a forged update causes harm, your company's name could be in headlines for the wrong reasons. Customers and partners would naturally **lose confidence**. As the dotmagazine article noted, digital trust is critical to customer loyalty and brand protection. A breach that can be tied to "outdated encryption" might make the board look negligent and the brand look ill-prepared. In competitive markets, being seen as insecure can drive customers to rivals. On the flip side, being proactive about quantum-safe security can be a positive differentiator (more on that in the next chapter).
- **Legal Liabilities:** Beyond regulatory fines, companies could face direct legal action. If confidential client data or communications (that were contractually expected to be kept secure) become compromised, clients may sue for damages. Directors and officers might be questioned on whether they fulfilled their fiduciary duty to anticipate foreseeable risks (and leading experts have indeed been warning of the quantum risk as a foreseeable development). It's not hard to envision class actions or derivative suits in the wake of a quantum-related security failure, alleging lack of proper risk management.

In essence, a failure of cryptography undermines **digital trust** – and without trust, digital business cannot function. The impact cascades from technical chaos to strategic catastrophe, touching finances, legal standing, customer relations, and even national security, depending on the organization. One study by KPMG summarized it well: a breach enabled by quantum could have "**catastrophic financial, reputational and legal consequences**" for organizations across sectors. The prudent question for executives is not *"Can we afford to address this now?"* but rather *"What will it cost us if we fail to address it?"* The **cost of inaction** may far exceed the investment needed to mitigate this risk.

From Urgency to Action: Preparing for the Quantum Era

By now it should be clear: **quantum computing represents a**

strategic risk to every digital enterprise's foundation of trust. This is not just an IT department issue; it's a boardroom issue. Much like climate risk or supply chain risk, it demands forward-looking governance. Business leaders need to signal urgency and set the tone: **prepare now to be safe later.** As one executive advisor put it, *"There is little time to lose for organizations to gain a deeper understanding of the risks quantum may pose...consider the lifetime value of their data and the impact of it being misused by bad actors."* In other words, align your security strategy with the reality that **the rules of encryption are changing.**

Thankfully, this is not all doom and gloom. The same global community of scientists and engineers that built our current cryptography has been racing to develop **quantum-resistant (post-quantum) cryptographic algorithms.** Standards bodies like NIST have already published new encryption and digital signature algorithms designed to withstand quantum attacks. These **PQC (Post-Quantum Cryptography)** solutions are the next-generation locks and keys for the quantum age. But deploying them will take time – often estimated at 5-10 years for a full transition in large organizations. That is why **planning and acting now** is critical. In the next chapter, we will dive into how organizations can **be safe** and stay ahead: building **crypto-agility** (the ability to rapidly swap out cryptographic systems), rolling out PQC across your IT landscape, and using **hybrid cryptography** (combining classical and quantum-safe methods) to stay resilient during the transition.

The key takeaway for executives is one of **proactive leadership.** A quantum-ready strategy is not just about avoiding threats; it's about maintaining the trust of customers, partners, and stakeholders in a changing technological landscape. PART III will outline how to turn this looming risk into an opportunity – by **being ahead, being safe, and ultimately winning** in the quantum era through prudent action and innovation. Remember: the companies that **prepare today** will be the trusted brands of tomorrow's digital economy, while those that delay may find their **digital foundations crumbling** when the quantum future arrives.

4
STANDARDS, TIMELINES, AND SIGNALS

Quantum computing is forcing a paradigm shift in cybersecurity: new cryptographic standards and global mandates are emerging to ensure organizations become "quantum-ready."

In boardrooms and executive suites worldwide, a clear signal is sounding: **the race is on to upgrade encryption before quantum computers arrive**. Governments and standards bodies are moving from awareness to action, publishing *post-quantum cryptography (PQC)* standards and setting hard deadlines for migration. This chapter provides a high-level roadmap of what's real – the new PQC standards, the mandates with specific timelines, and the readiness signals executives should heed. The tone is sharp and grounded, focusing on actionable facts rather than technical minutiae. The message: **proactive planning and early moves will be rewarded**, while delay only increases exposure.

NIST's Post-Quantum Standards and Timeline

The U.S. National Institute of Standards and Technology (NIST) has now formalized the first wave of quantum-resistant cryptographic algorithms, after an eight-year worldwide competition. In August 2024, NIST published three Federal Information Processing Standards (FIPS) that will replace today's vulnerable public-key algorithms (like RSA and ECC) in the quantum era. These include one encryption/key-establishment method and two digital signature schemes:

- **FIPS 203 (ML-KEM)** – Based on the CRYSTALS-Kyber algorithm, this **Module-Lattice Key Encapsulation Mechanism** is the new primary standard for encrypting data and establishing secure keys. Kyber (ML-KEM) offers strong security with relatively small keys and fast performance, making it a go-to replacement for RSA key exchange.

- **FIPS 204 (ML-DSA)** – Based on the CRYSTALS-Dilithium algorithm, this **Module-Lattice Digital Signature Algorithm** is the main standard for digital signatures (e.g. signing software, documents, certificates). Dilithium (ML-DSA) provides efficient signing and verification with strong security, and is expected to see broad use in software signing and authentication.
- **FIPS 205 (SLH-DSA)** – Based on the SPHINCS+ algorithm, this **Stateless Hash-based Digital Signature Algorithm** provides an *alternate* signature method. SPHINCS+ is slower and produces larger signatures than Dilithium, but it relies on hash functions (not lattices) and serves as a conservative backup in case lattice-based schemes are ever weakened. This diversity ensures no single point of failure in our crypto toolbox.

Together, these three standards mark a milestone: **the first quantum-safe encryption and signature schemes ready for use in mainstream technology**. NIST is urging organizations to begin integrating them *now*, recognizing that full deployment will take time. Dustin Moody, NIST's PQC program lead, put it plainly: *"We encourage system administrators to start integrating them into their systems immediately, because full integration will take time."*

NIST isn't stopping at the initial trio. A **fourth algorithm** – an additional digital signature scheme called FALCON – is also slated for standardization (as FIPS 206) by late 2024, to offer a second lattice-based signature option for specialized use cases. More importantly, NIST has planned **backup encryption algorithms** to hedge against any future cracks in the primary scheme. In March 2025, NIST **selected the HQC algorithm** (Hamming Quasi-Cyclic) as a **backup** for general encryption. HQC is based on a different mathematical foundation (error-correcting codes instead of lattices), providing a second line of defense if a breakthrough ever compromises Kyber.

Executive Signal: Even in the unlikely event that our first-choice quantum-safe *encryption (Kyber) had a flaw, NIST is ensuring there's a Plan B.*

A draft standard for HQC will be published in 2026, with finalization expected in 2027. This means by 2027 we will have at least **five** NIST-approved PQC algorithms (Kyber + HQC for encryption; Dilithium, FALCON, SPHINCS+ for signatures) covering all key cryptographic needs.

Timeline context: NIST's roadmap is essentially on track. The initial PQC standards were finalized in 2024, roughly eight years after the effort began. The backup encryption standard (HQC) will follow by 2027. For executives, this translates to a clear window: **the standards are here** or imminent – the next 2–3 years will see certified implementations rolling out, and by **2025–2027** the technical foundation for quantum-safe cryptography will be fully in place. There is no reason to wait; the era of PQC has effectively begun.

NSA's CNSA 2.0: National Security Mandates

While NIST handles the science of new algorithms, the U.S. National Security Agency (NSA) is driving policy for high-security systems. In September 2022, NSA announced **CNSA 2.0**, the updated Suite of **Commercial National Security Algorithms** required for U.S. National Security Systems (systems that handle classified or military communications). CNSA 2.0 explicitly embraces the coming quantum-resistant algorithms. It adopts the NIST selections (e.g. lattice-based encryption and signatures) as the core of its new suite, while retaining robust symmetric algorithms like AES-256 and SHA-384 that remain secure against quantum attacks.

Mandate and timeline: The NSA has made it clear that **all U.S. national security systems must transition to quantum-resistant cryptography "as soon as possible" – with an ultimate deadline of 2035**. In other words, any communications or data meant to stay secure past 2035 **must be protected by CNSA 2.0 algorithms**. To meet this, NSA's roadmap sets phased objectives. For example, **software and firmware signing** used in defense and intelligence communities should start switching to quantum-safe signatures immediately, **using only CNSA 2.0 algorithms by 2030**. Other categories have similar targets: critical web and network systems must prefer PQC by the mid-2020s and exclusively use them by 2033. By 2035, the expectation is that **everything in national security – from secure communications links to stored classified data – will be using quantum-resistant encryption and authentication**.

These mandates are more than guidelines – they are requirements that drive procurement and budgets. The NSA has authority over national security systems, and CNSA 2.0 compliance will be enforced.

Executive Signal: If your organization works with U.S. national security (e.g. *as a contractor or technology supplier), the clock is already ticking. CNSA 2.0 is effectively a* ***hard deadline for 2035****, and interim milestones (like 2030 for code-signing systems) mean quantum-safe upgrades must be in your 5-year IT plans.*

Beyond dates, NSA's policy sends a strong **signal to the market**. By declaring the specific algorithms and a timeline, NSA is pushing vendors to develop compliant products (from VPNs and radios to software update mechanisms) *now*. They have also cautioned not to implement unproven algorithms too early – *wait for NIST standards and validated implementations* – but **do not wait to prepare budgets and transition plans**. The national security community knows this transition will be complex and is coordinating closely with industry. The bottom line: **quantum-safe crypto is becoming mandatory for the most sensitive systems**, and that mandate is cascading into the broader technology ecosystem via government purchasing power.

U.S. Federal Deadlines: NSM-10 and OMB Guidance

It's not just the intelligence and defense world gearing up. The **entire U.S. federal government** is under orders to begin migrating to post-quantum cryptography. President Biden's National Security Memorandum 10 (NSM-10), issued May 2022, set the stage by stating that "the United States must prioritize the timely and equitable transition of cryptographic systems to quantum-resistant cryptography… with the goal of mitigating as much of the quantum risk as feasible by 2035." This high-level directive was soon followed by concrete marching orders from the Office of Management and Budget (OMB).

In November 2022, OMB released **Memo M-23-02, "Migrating to Post-Quantum Cryptography,"** which requires federal agencies to **take inventory of their cryptographic systems and assets** as a first step. The memo mandates that by **May 4, 2023**, every agency had to submit a prioritized inventory of all systems relying on cryptography, and identify any particularly sensitive data that could be at risk from quantum decryption. This inventory must be updated and reported annually **through 2035**. Why 2035? It aligns with the White House's target date by which the quantum threat should be largely addressed. Essentially, the U.S. government is **working backwards from 2035**, making sure progress is tracked year by year.

Crucially, OMB's guidance didn't stop at inventory. Agencies were instructed to **establish plans for transitioning to PQC**, including testing of new algorithms once available, and to **prioritize funding** for this in their budgets. There is also an implicit expectation of **"crypto-agility"** – systems should be designed to be upgradable with new algorithms when needed. (While not all agency systems are agile today, future procurements will likely require it, to avoid being stuck with non-upgradeable crypto.) By early 2024, each agency was expected to submit a full strategy for moving its systems to quantum-resistant encryption, per OMB and the Cybersecurity & Infrastructure Security Agency (CISA) guidelines.

For private-sector leaders, these federal mandates serve as a **valuable blueprint**. They highlight best practices that any large organization can adopt: perform a **crypto inventory** (know where and how your data is protected), assess **data sensitivity and shelf-life** (which data, if stolen today and decrypted in 5–10 years, would cause harm?), and develop a **migration roadmap**. The U.S. government is essentially saying that *by 2025–2026, every organization should know its cryptographic exposure and have a plan to swap in PQC algorithms.*

Executive Signal: *Even if you're not compelled by law, treating 2025–26 as your deadline to complete a crypto inventory and PQC transition plan is a wise move.*

The federal government's house will be in order; regulators and cyber insurers will naturally start asking if yours is too.

Finally, note that in late 2022 Congress passed the **Quantum Computing Cybersecurity Preparedness Act**, which reinforces some of these requirements and transparency. It directs OMB to report on progress and ensures Congress is kept informed of agency readiness. This adds pressure and accountability. From an executive standpoint, consider that *any business dealing with federal agencies or critical infrastructure could face similar scrutiny.* Being quantum-ready could become a market differentiator – or simply a requirement to do business – in the near future.

Global Quantum-Ready Roadmaps: UK, EU, and APAC

The push for quantum-resistant security is global. Allies and economic blocs around the world are issuing their own guidance and mandates, often aligning with the U.S. 2035 horizon while sometimes pushing for even earlier action. Here's a quick tour:

United Kingdom: Phased Migration by 2035

The UK's National Cyber Security Centre (NCSC) has laid out a clear, phased plan for all British organizations to become quantum-safe by 2035. In March 2025, the NCSC published migration timelines with **key milestones**:

- **By 2028:** Complete a *discovery and planning phase*. Organizations should *identify all systems and services using cryptography* and draft an initial PQC migration plan. (In simpler terms: know where your encryption is and decide what needs fixing first.)
- **By 2031:** Have all *high-priority systems* transitioned or upgraded with quantum-safe solutions, and refine the roadmap for everything else. This means critical infrastructure, sensitive data systems, and any exposed public-facing encryption (like customer portals) should be on PQC by this date. Plans should be concrete for the remaining lower-priority areas.
- **By 2035: Complete the migration** across *all* systems, services, and products. At this point, traditional public-key crypto (RSA/ECC) should be fully replaced or wrapped with PQC in the UK's digital ecosystem. Any stragglers or hard-to-upgrade legacy tech must be dealt with by this "long stop" date.

The NCSC emphasizes these deadlines are **indicative but important** – they set a national expectation. They also note that some legacy or niche technologies might struggle to meet 2035, but **all organizations should work toward those dates** as firmly as possible. The rationale is straightforward: a ten-year window (2025 to 2035) is generally the minimum needed to enact such a sweeping change across government and industry. The UK plan also smartly highlights *early planning and vendor engagement*. Businesses are urged to coordinate with suppliers and start testing PQC solutions well before the final deadlines. Waiting until 2030 to start implementation, for example, would risk a last-minute crunch or security gap.

European Union: Coordinated EU-Wide Timeline

Across the Channel, the European Union is moving in concert. In April 2024, the European Commission, together with EU member states, issued a **"Coordinated Implementation Roadmap"** for PQC.

This roadmap calls on all member states to **begin transitioning to post-quantum cryptography by the end of 2026**. In practice, that means EU governments and critical industries should start upgrading systems or at least piloting PQC solutions no later than 2026.

A notable EU target is that **critical infrastructure protections should be quantum-resistant by end of 2030**. Sectors like energy, finance, healthcare, transportation – many covered under the EU's NIS2 Directive for network and information security – are expected to prioritize PQC so that the most essential services do not rely on breakable crypto beyond 2030. The EU's approach is somewhat aggressive: essentially **"start now, finish the most critical pieces by 2030."** We can infer that full migration for less critical systems would follow the global trend into the early 2030s, likely landing around 2035 as well. Indeed, EU officials have openly stated that the goal is to keep pace with allies and that the **2035 horizon for complete transition is a guiding benchmark** (though individual member state plans may vary slightly).

Coordination is key in the EU strategy. The roadmap recommends each member state develop national PQC migration plans in sync, share progress via the NIS Cooperation Group, and jointly raise awareness across industries. For executives operating in Europe, this means you can expect **regulatory pressure** in the coming years to demonstrate quantum-readiness. Whether through updates to NIS2 implementation, new ENISA (EU cybersecurity agency) guidelines, or sector-specific regulations, the EU is signaling that *quantum-safe crypto is part of "state of the art security,"* which is a legal requirement under EU law for protecting data. As a practical step, businesses in the EU should follow the Commission's recommendation: *inventory your cryptographic systems by 2025–2026 and devise a plan so that by 2030 your critical data is safe, and by 2035 all your data is safe.* The timeline is ambitious but in line with the urgency of the threat.

Asia-Pacific: Early Signals and Ambitious Targets

Several Asia-Pacific nations are also acting decisively:

- **Australia** has been especially proactive. The Australian Cyber Security Centre (ACSC) published guidance in 2023 on "Planning for Post-Quantum Cryptography," urging organizations to start identifying crypto dependencies and to ensure new systems are crypto-agile. In 2024, Australian regulators went a step further by proposing to **phase out**

certain "weak" encryption algorithms by 2030, effectively *five years ahead* of the U.S. NIST 2035 schedule. In other words, Australia is eyeing 2030 as the drop-dead date after which outdated algorithms like RSA-2048 or smaller ECC should no longer be in use for sensitive applications. This aggressive stance reflects a desire to stay ahead of the threat and echoes Australia's broader cybersecurity strategy of moving fast on emerging risks.

- **Japan** has been steadily preparing as well. The government (through CRYPTREC, its cryptographic evaluation committee) issued PQC guidelines as early as 2022, closely monitoring NIST's competition and encouraging domestic organizations to start planning transitions. Japan is aligning its selection of algorithms with the global consensus (Kyber, etc.), and while no hard national deadline has been announced, the expectation is that Japan will transition on a similar 10–year timeline. Japanese industries (automotive, finance, electronics) are heavily involved in global supply chains, so their push for crypto-agility and PQC support is strong. We can expect formal Japanese government roadmaps to coalesce once NIST standards are fully finalized, if not sooner.
- **South Korea** has launched a national PQC project including local algorithm development (the "Korea PQC" algorithms) and a roadmap aiming for **complete PQC adoption by 2035**, with pilot migrations in 2025–2028. This mirrors the U.S./UK approach. The inclusion of domestic algorithms also highlights a trend: some countries want their own options (for sovereignty reasons) alongside NIST's standards. But the end goal is the same – quantum-safe encryption deployed nation-wide by the mid-2030s.
- **Singapore** stands out as a financial and tech hub taking early action. The Cyber Security Agency of Singapore (CSA) announced it will roll out **quantum-security guidelines starting in 2025** to help organizations prepare. They are prioritizing essential sectors like finance, healthcare, telecom, and government services. Singaporean experts acknowledge the transition could take a decade or more, and local businesses are advised to assess the "value and lifespan" of their data now – if information needs to remain confidential into the mid-2030s, it should be protected with quantum-safe

methods sooner rather than later. The Singapore approach underscores a practical point: *not all data is equal.* For example, a routine e-commerce transaction today might not need PQC if it's only sensitive for a year, but a health record or intellectual property with a 20-year importance absolutely does. Executives in all regions should make this kind of data triage part of their planning.

In summary, the international picture shows a **convergence around the early-to-mid 2030s** as the timeframe by which quantum-vulnerable cryptography should be retired. Some jurisdictions aim to beat that by a few years for critical systems (EU by 2030, Australia by 2030, UK high-priority by 2031). Others are right on the mark of 2035 (U.S. national security, UK final deadline, South Korea). The key point for a global company is clear: **no matter where you operate, the next 5–10 years will bring increasing mandates to inventory, test, and swap out your encryption and authentication mechanisms.** Being a laggard could mean being out of compliance in multiple markets – or worse, being the weak link in a future supply chain security audit.

Protocols, Products, and Crypto-Agility

Standards and mandates set the destination; now we consider the vehicle for getting there. Implementing PQC is not just a checkbox upgrade – it will require updates deep in the digital plumbing of networks and applications. Two concepts are paramount: **crypto-agility** and *hybrid solutions.*

Crypto-agility is the capability to *swap out cryptographic algorithms with minimal disruption.* In a world where algorithms can become obsolete (as quantum computing threatens to do to RSA/ECC), agility is essentially an insurance policy. It means designing systems so that cryptographic modules (like TLS libraries, VPN clients, authentication systems) support multiple algorithms and can be updated through configuration or patches, rather than a complete overhaul. Many regulators are explicitly calling for crypto-agility. For instance, in critical infrastructure sectors the ability to "swap out algorithms easily" is seen as crucial to facilitate the quantum transition. It ensures that if one PQC algorithm falters or new ones emerge, organizations can adapt quickly without massive re-engineering.

Executive Signal: *Ask your CTO/CIO: "How quickly can we change the cryptography underpinning our systems if we need to?" If the answer is in months or years (or blank stares), that's a red flag.*

Modern system design, especially for anything touching secure communications, should incorporate agile crypto frameworks (e.g. supporting TLS cipher suite updates, pluggable cryptographic providers, and centralized key management that can handle new algorithms).

Industry is responding. Forward-looking tech companies and open-source projects have been building agility into their products. One open-source security engineer quipped: *"Let's champion crypto-agility in everything we create — ensuring algorithms, keys, device identities, and roots of trust are all updatable. Remember, this won't be the last time we need to manage migrations."*. In practice, this means when you procure software or hardware, **make PQC support and crypto-agility a selection criterion**. If a vendor can't articulate how they will add PQC algorithms or if their solution has hard-coded cryptography that can't be easily changed, consider that technical debt. Many enterprise security products (VPNs, databases, IoT devices) will require firmware or software updates to add PQC – check that your suppliers have these on their roadmap.

In parallel, the **Internet Engineering Task Force (IETF)** – which governs protocols like TLS (for web encryption) and IPsec/IKE (for VPNs) – has been developing standards for *hybrid cryptographic modes*. A hybrid approach means using a combination of a classic algorithm and a PQC algorithm **together** during a transition period. This "belt-and-suspenders" method is recommended by experts. For example, there are IETF drafts for **TLS 1.3** that define a hybrid key exchange: your browser and server could negotiate both an elliptic-curve Diffie-Hellman key exchange *and* a Kyber key encapsulation, and use both keys to derive the session secret. Even if one algorithm is later broken, the session stays secure as long as the other holds. These hybrid modes are already being tested: major browser vendors and tech companies have run trial deployments (e.g. Chrome and Firefox experimented with a hybrid X25519+Kyber cipher suite). Similarly, for VPNs, the IETF is working on hybrid authentication for IKEv2 (the handshake behind IPsec) allowing, say, an RSA certificate to be paired with a Dilithium signature in authentication.

What this means for executives is that **standards bodies are smoothing the implementation path**. You likely won't have to choose a "flag day" to flip everything to PQC in one go. Instead, best practice will be to *enable hybrid cryptography in the 2020s*, gradually introduce PQC alongside existing encryption, and then phase out the older algorithms by the 2030s deadline. Many commercial products will deliver these capabilities via updates: e.g. an update to your TLS stack to support a hybrid cipher suite, which you can turn on via configuration. Ensure your technical teams are aware of these developments – a crypto-agile system can deploy such updates as they become available.

One more element of agility is **protocol flexibility**. Some older protocols might not have supported new algorithms easily (for example, hard-coded assumptions about key sizes). The move to TLS 1.3, modern IPsec, and quantum-safe VPN protocols will be part of the journey. If you're running very old protocol versions in your environment, plan to upgrade. This is analogous to past transitions (like moving from older SSL versions to TLS 1.2/1.3 for better security). The good news: TLS 1.3 is already designed with algorithm agility in mind and will be the foundation for quantum-safe web communications. So, an *action item for CIOs* might be: audit where your organization still uses outdated or non-agile protocols and make upgrading them part of your PQC readiness plan.

The Quantum-Ready Timeline: Executive Actions Now

Bringing it all together, here is a *bullet-style timeline* of key milestones and deadlines on the road to quantum resilience, and what executives should be doing at each stage:

- **2024 – First PQC standards finalized:** NIST publishes FIPS 203/204/205 (Kyber, Dilithium, SPHINCS+) as official standards.

  Action: Begin evaluating these algorithms in pilot projects; update vendor requirements to include PQC support.

- **2025 – Mandates kick in:** U.S. federal agencies have completed initial crypto inventories; NIST selects HQC as backup encryption algorithm.

  Action: Ensure your organization has a cryptography inventory project underway (if not completed). Pressure-test your business against a "quantum day" – what data would be at risk if a quantum computer appeared in 5 years?

- **By End of 2026 – Global commencement:** EU requires all member states to **start** PQC transition by this date.

  Action: Treat 2026 as your target to have migration plans in place and initial quantum-safe solutions deployed in test environments. Also, any new system deployed from now on should be crypto-agile by design.

- **2028 – UK discovery phase deadline:** Per NCSC, have completed crypto discovery and drafted migration plans by this year. (By this time, major vendors will offer PQC-enabled products).

  Action: Aim to finish your organization's full cryptographic asset inventory no later than 2028. Identify systems that will be hardest to upgrade and allocate budget/resources to them first.

- **2030 – Quantum-safe critical infrastructure:** EU's deadline for critical sectors (energy, finance, etc.) to have PQC in place. Also, NSA expects all *software and firmware signing* for national security systems to use only quantum-resistant algorithms by this year. Australia plans to **deprecate legacy encryption by 2030** as well.

  Action: Set 2030 as the target for your most critical assets – customer-facing services, crown-jewel data, and crucial IP – to be quantum-safe. Even if not legally required, it's a sound business deadline to protect core assets and maintain trust.

- **2031 – UK high-priority systems migrated:** NCSC expects early migration of highest-priority systems complete by 2031.

  Action: By this point, ensure an *enterprise-wide PQC roll-out* is actively in progress. Your quarterly cyber risk reports to the board should show significant reduction of legacy crypto usage year-over-year.

- **2033 – Broader systems cutover:** U.S. NSA's interim milestones indicate that by 2033, most networking equipment and operating systems in NSS environments should use exclusively PQC algorithms.

  Action: Target 2023–2033 as your main execution window for full migration. Any systems still not upgraded by 2033 should be isolated or risk-managed heavily, as time is nearly up.

- **2035 – "Q-Day" readiness achieved:** U.S. National Security Systems must be fully quantum-resistant. UK's final deadline for all systems across sectors is also 2035. Many experts predict that by around 2035, either a cryptanalytically relevant quantum computer could exist or we must assume it *could* exist.

Action: *Mission accomplished.* By 2035, your organization should have completed its crypto transition. Treat any use of old algorithms past this date as an emergency vulnerability. Your focus can shift to tightening cyber resilience with the new tools (and monitoring advances in case new threats emerge).

Throughout this timeline, one thing is constant: **delaying increases risk**. The oft-cited concern is *"Harvest Now, Decrypt Later."* Adversaries can steal encrypted data today and simply hold it, waiting for the day a quantum computer can decrypt it. Every year of delay is an additional year of data that might be secretly captured and later exposed. Particularly for data with long sensitivity (think healthcare records, personal identifiable information, trade secrets, military secrets), a breach in 2025 could turn into a plaintext compromise in 2035 if not protected by quantum-safe crypto. That looming threat is what's driving these seemingly distant deadlines into near-term action items.

Executive Signal: This is a rare cybersecurity challenge with a *predictable* deadline.

We know roughly when we need to be done (by 2035, if not earlier), and we have the tools emerging now to meet it. Winning organizations will be those that **act early** – conducting crypto audits, pressuring vendors for solutions, and fostering a culture of crypto-agility. By treating quantum readiness as a strategic initiative (on par with digital transformation or cloud migration), you not only mitigate the quantum risk but often improve your overall cyber posture (through better inventory, updated systems, and stronger cryptographic governance).

What to do now:

- **Launch your cryptography inventory and risk assessment** (if you haven't already). Know your exposure and identify high-risk areas where "encrypt now, decrypt later" is a concern. Aim to have this completed by 2025–2026, aligning with government guidelines.
- **Incorporate PQC into your strategic planning and budgets.** Treat crypto upgrades as a multi-year project (3–7 years for most large enterprises). This isn't a simple patch; it may involve upgrading applications, hardware (like HSMs or TPMs), and protocols. Allocate funding and personnel

accordingly.

- **Engage vendors and require quantum-safe roadmaps.** When negotiating contracts or renewals with IT and cloud providers, ask about their support for PQC algorithms and crypto-agility. If a vendor's product is critical for you (e.g. a core database or network appliance), you need assurances it will have a quantum-safe upgrade well before 2030. Regulators are already hinting that *vendor readiness* is part of the equation for critical sectors.
- **Build crypto-agility into everything.** As mentioned, any new system design should assume that algorithms might need to change. Leverage architectures and libraries that allow easy swap-out of crypto components. This also future-proofs you against other cryptographic developments.
- **Start with hybrid solutions where available.** For instance, experiment with enabling a hybrid TLS mode in a test environment or for internal services once standards mature. This lets you vet PQC performance and compatibility without fully committing to it alone. It's a smart way to gradually introduce the technology.
- **Educate and train your teams.** Make sure your CIO, CISO, and technical staff are up to date on PQC developments. Identify knowledge gaps – cryptographic expertise is suddenly at a premium again. You may need to bring in consultants or allocate training so that your organization isn't flying blind in implementing these new algorithms.

In closing, achieving **quantum advantage** in security – being ahead, being safe, winning the trust of customers and stakeholders – is doable if you heed the standards and mandates now in plain view. This is not theoretical policy on a distant horizon; it's a concrete timeline unfolding today. With coordinated effort, the looming quantum disruption can be met with confidence. The organizations that move decisively will not only neutralize the quantum threat, they will reinforce their security foundations for whatever comes next. Now is the time to be *quantum-ready*.

PART III
THE ADVANTAGE FRAMEWORK

This is the core playbook. It lays out the three mandates: **Be Ahead, Be Safe, Win.** It also provides the strategies and practices leaders can adopt now to build resilience, seize opportunity, and turn readiness into competitive edge.

BE AHEAD, BE SAFE, WIN

WIN

Leverage quantum opportunities

QUANTUM ADVANTAGE

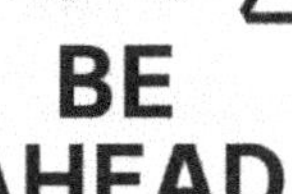

BE AHEAD

Assess readiness
Scan horizon

BE SAFE

Implement quantum-safe security
Ensure resilience

5 BE SAFE: THE QUANTUM RESILIENCE PLAYBOOK

In the face of quantum threats, **being safe** is not about panicking or over-hyping worst-case scenarios. It's about proactive leadership in digital trust and resilience. Boards and CXOs must champion concrete steps now to protect their organization's "trust layer" – the encryption, identity, and data security foundations that could be broken by quantum computing. Chapter 3 underscored the urgency of the quantum risk; this chapter translates that urgency into an actionable playbook. By embedding quantum resilience into your strategy, you demonstrate not fear, but foresight and trustworthiness. The goal is clear: ensure your organization can **withstand and adapt** to cryptographic upheavals, maintain the confidence of customers and partners, and emerge stronger. This playbook outlines how to get there.

Crypto-Agility: The Foundation of Quantum Resilience

Crypto-agility is the cornerstone of being safe in the quantum era. Cryptographic agility means your systems can **rapidly swap out or upgrade encryption algorithms** without major disruption. Think of it as future-proofing your security: if a cryptographic method is weakened – whether by quantum computers or any new vulnerability – a crypto-agile organization can pivot quickly, updating keys and algorithms across all systems with minimal downtime. In practice, crypto-agility involves building flexibility into your infrastructure and processes so that encryption isn't "hardwired" in ways that are costly or slow to change.

From a board perspective, crypto-agility is no longer a niche technical topic; it's a business continuity issue. **Executives should be asking:**

"How fast could we update our encryption if we had to? Do we know everywhere it's used? Who is responsible for overseeing this?" These questions bring crypto-agility into governance view. For example, consider digital certificates that secure your websites, VPNs, and software updates – can your team revoke and replace those certificates across the enterprise on short notice if an algorithm is compromised? If the answer is "not sure" or "it would take a while," that's a red flag.

Executive Signal: If your IT systems can't rotate keys without downtime, you're not crypto-agile.

Leading organizations treat crypto-agility as a design principle, not an afterthought. This means **building encryption into systems in a modular way**, using updatable libraries and protocols, and decoupling cryptographic components from business logic where possible. It also means establishing policies and drills for crypto updates – much like disaster recovery exercises, you should have *cryptographic contingency plans.* For instance, some companies conduct "crypto fire drills" where they simulate the sudden deprecation of an algorithm and practice deploying a new one. Boards should insist on such preparedness. A crypto-agile posture not only reduces technical risk, but also **streamlines compliance** with evolving regulations and standards. It can even lower long-term costs by avoiding expensive last-minute overhauls.

Most importantly, crypto-agility buys you time and resiliency in the quantum race. It's the **antidote to cryptographic complacency**. As one industry report put it, the goal of crypto-agility is simple: *"to enable business continuity if/when existing cryptography is compromised or weakened"*. In short, it's the capability that ensures you won't be stuck on yesterday's security while competitors (or adversaries) race ahead.

Key questions for leadership:

- *Do we have an inventory of all cryptography in use (algorithms, keys, certificates, libraries)?*
- *Can we change our encryption algorithms across systems* ***within days or weeks****, not months or years?*
- *Who in our organization is accountable for cryptographic resilience and updates?*
- *Are we training our IT teams to be "crypto-agile," and do we have playbooks for rapid crypto replacement?*

By asking these questions, boards set the tone that crypto-agility is a priority. It signals to management that quantum resilience isn't just an IT project – it's a strategic imperative touching risk management, operations, and trust. In many cases, answering these questions will reveal gaps that need addressing immediately.

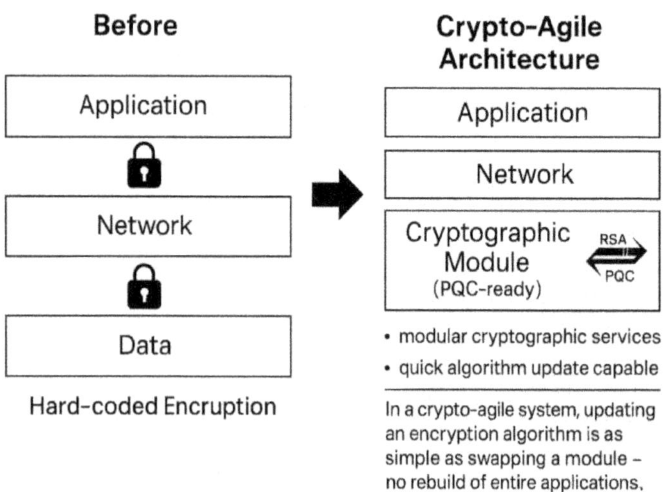

Post-Quantum Cryptography (PQC): New Standards, New Strategy

Becoming quantum-safe starts with adopting **post-quantum cryptography (PQC)** – the new generation of encryption algorithms designed to withstand attacks from quantum computers. The good news is that PQC is here **today**. After a multi-year international effort, standards bodies have introduced robust algorithms to replace our vulnerable RSA and ECC (elliptic-curve) cryptosystems. In August 2024, the U.S. National Institute of Standards and Technology (NIST) published its first set of **quantum-resistant algorithms** as official standards (FIPS 203, 204, 205), including **CRYSTALS-Kyber** for encryption/key-establishment and **CRYSTALS-Dilithium** for digital signatures. These were the winners of NIST's rigorous PQC competition and are **explicitly engineered to resist** the known quantum attacks that would break classical cryptography. More PQC algorithms are on the way (for example, NIST selected an additional encryption scheme, HQC, in 2025) to broaden the toolbox.

What does this timeline mean for executives? It means the horse has left the barn – the **standards are available**, and the transition period is already underway. Governments are moving: the White House estimates over **$7 billion** will be spent in the next decade to remediate and upgrade federal cryptography for quantum safety. Regulators and industry groups are sounding the alarm that **"start the upgrade process now"**. In other words, the question is no longer *if* PQC should be adopted, but **when and how**. Forward-looking companies have begun experimenting with PQC in their environments, and some tech platforms (browsers, VPN solutions, etc.) are shipping initial support. However, adoption in the broader industry is still nascent – surveys show that as of 2024, only about **7%** of U.S. federal agencies even had a formal PQC transition plan in place. That gap between awareness and action is your opportunity to lead.

For executive teams, **PQC adoption needs to be framed as a strategic upgrade** to the company's security infrastructure. Just as you budget for cloud transformation or AI, you must budget for quantum-resistant security. The transition will span multiple years; NIST and national cybersecurity centers project that by around **2030** we should be using PQC widely (for example, NIST's goal is that by 2030, "all TLS handshakes should be quantum-safe" in new standards). That may feel far off in board-year terms, but the intervening steps – evaluation, pilot projects, procurement updates, talent training – need to happen **now** to meet that timeline. Remember that **past cryptographic transitions have taken a decade or more**, even without the pressure of a new threat. The shift from SHA-1 to SHA-256, or from 3DES to AES, dragged on because of the complexity of replacing algorithms across legacy systems. We are facing a similar multi-year journey for PQC, and likely an even more complex one given the ubiquity of cryptography in modern IT.

What should leaders do about PQC? First, ensure your technology teams are **testing the new algorithms** now. For instance, have them evaluate the performance of Kyber and Dilithium in your use cases – PQC algorithms often have larger key sizes or signature sizes and different computational profiles than RSA/ECC. Knowing the impact on network latency or CPU load is crucial for planning. (The encouraging news: early experiments found that using a PQC algorithm in a TLS handshake added only a few milliseconds of overhead , though impacts can vary by context.) Second, **set a target timeline** for when your organization will be quantum-ready. Many companies are publicly committing to being quantum-safe by, say, 2030, aligning with anticipated regulatory expectations. Internally, that timeline will drive your roadmap priorities (discussed below). Third, **foster a culture of cryptographic vigilance**. Make sure your CISO and security architects are tracking PQC developments: subscribe to NIST updates, join industry working groups, and monitor guidance from agencies like DHS, NSA, and ENISA. PQC standards will continue to evolve (e.g. new algorithms, updates to parameters) and being plugged into the community ensures you won't be caught off-guard.

Finally, **communicate the strategy**. Boards should expect management to articulate a clear plan: "Here is how we will transition our encryption to PQC over the next 3–5–7 years, and here are the milestones." This plan becomes part of your cyber risk oversight. By publicly acknowledging quantum threats and your proactive response, you also send a signal to stakeholders that your organization values trust and resilience. As we'll explore later, being ahead on security can become a market differentiator.

Hybrid Cryptography: Bridging Today and Tomorrow

One practical strategy to "be safe" during the PQC migration is **hybrid cryptography** – using both classical and quantum-safe methods in tandem. Hybrid cryptography is essentially a **belt-and-suspenders approach** to encryption: you combine a tried-and-true classical algorithm with a new PQC algorithm so that a message or connection remains secure unless *both* algorithms are broken. This approach addresses two challenges: (1) PQC algorithms are new and haven't undergone decades of real-world testing (so we hedge against any unforeseen weakness by also using classical crypto), and (2) not everyone can or will upgrade at once (so the classical part maintains compatibility with legacy systems while the PQC part adds quantum protection).

In technical terms, a hybrid encryption scheme might do something like this: when establishing a secure connection, the system performs **two key exchanges** – one using a classical algorithm (e.g. an elliptic-curve Diffie-Hellman exchange) and one using a post-quantum algorithm (e.g. Kyber). The two resulting keys are then combined (mixed) into a single session key that is used to encrypt the communication. An eavesdropper would need to break **both** the classical and the quantum-safe exchange to derive the key, meaning the connection is safe unless quantum (and classical) attacks succeed simultaneously. Similarly, for digital signatures, one could require a signature that is verifiable with both an RSA/ECC key **and** a PQC signature like Dilithium – the message is considered authentic only if both signatures check out.

Why use hybrids? Because they enable a gradual, *risk-managed transition.* For example, consider the challenge of upgrading a global banking system to PQC. Not all client applications, partner systems, or devices will support Dilithium or Kyber from day one. With hybrid modes, you don't have to flip a switch overnight. You can introduce PQC alongside classical crypto: updated systems will get the added quantum safety, while older systems can continue operating with the classical algorithms they understand. Standards bodies have embraced this: the Internet Engineering Task Force (IETF) has already published standards like **RFC 8784** for hybrid key exchange in IPsec/IKEv2 VPN protocols, and is working on similar standards for TLS handshakes. These standards allow, for instance, a VPN tunnel to perform a parallel ECDH and Kyber exchange and use both keys, as described above. Major tech players have tested this concept; Google and Cloudflare's 2019 public experiment with a hybrid TLS 1.3 (combining X25519 elliptic-curve and a PQC algorithm) demonstrated it could be done with minimal performance impact.

For executives, the takeaway is that hybrid cryptography provides a **pragmatic bridge** between today's infrastructure and tomorrow's. It lets you deploy quantum-safe measures **now** without stranding users or systems that aren't ready. In sectors like healthcare, finance, and critical infrastructure, where downtime or interoperability failure is unacceptable, this is a crucial advantage. Hybrid deployments can be an interim state for the next 5-10 years as PQC matures and proves itself. The NSA's recent guidance (CNSA 2.0 suite) in fact recommends hybrid modes for national security systems during this transition period – essentially double-encrypting with classical + PQC for high-assurance data.

A practical example: imagine a company's internal network where some applications have been upgraded to support PQC and others have not. Using a hybrid TLS solution, when those applications communicate, they execute both types of key exchange. Older endpoints will ignore the PQC part if they don't recognize it, but they will still complete the handshake with the classical part; newer endpoints will use both and get the extra security. In either case, the traffic is not left exclusively dependent on the old algorithms. **The bottom line** – hybrid crypto means you don't have to wait for 100% deployment of PQC to start getting protection, and you don't risk cutting off partners or customers who haven't upgraded. It's an *overlapping safety net.* Executives should ask their CISOs and CTOs: *"Are we exploring hybrid encryption options for our critical systems?"* and *"Do our VPN, TLS, and email security solutions support quantum-safe modes?"* If the answer is no, that's an area to invest in pilot projects (more on those next).

The Quantum-Safe Implementation Roadmap

Becoming quantum-resilient is a journey. This section lays out a high-level **roadmap** that organizations can follow – a progression from initial discovery to full deployment and governance. Each step is a play the leadership team should champion:

1. **Inventory Your Cryptography** – *"You can't protect what you don't know."* Start by mapping out **where and how cryptography is used** across the enterprise. This inventory is the foundation of everything to come. Many organizations are surprised by how widespread (and sometimes hidden) encryption is in their systems. Your inventory should catalog: all digital certificates and where they are deployed (web domains, internal services, mobile apps); all cryptographic libraries and modules in use; the algorithms and key lengths protecting data at rest and in transit; hardware cryptography (smart cards, Hardware Security Modules, TPMs); and third-party services that perform crypto on your behalf (cloud platforms, SaaS apps, payment gateways, etc.). Don't forget "embedded" cryptography – for example, an IoT device, database appliance, or VPN router might have encryption baked in that isn't immediately visible. This step will likely require a dedicated effort with tools and cross-team cooperation. The **board's role** here is to ensure management allocates resources to do this thoroughly. Until you have a clear cryptographic asset inventory, your quantum risk cannot be quantified or prioritized.

2. **Assess and Prioritize Risks** – Not all data and systems are equal. Once you know your cryptographic footprint, analyze which assets would cause the most damage if their encryption was broken. Identify your "crown jewels" – for example, customer PII databases, financial transaction records, sensitive intellectual property, authentication systems, etc. – and note

where they rely on vulnerable encryption. Also consider the **lifetime of the data**: information that needs to remain confidential for many years (medical records, secrets subject to long-term privacy laws, long-lived credentials like code-signing certificates) is a higher priority for quantum-safe protection than data that loses sensitivity after a short time. A crucial part of this assessment is the **harvest-now, decrypt-later** scenario : assume adversaries might be stealing encrypted data today to decrypt in a few years. Which data would hurt you most in that case? Those should be first in line for upgrading to PQC. This risk assessment phase informs a ranked list of where to apply fixes first. It's also the time to evaluate **operational constraints** – for instance, if a critical system can't easily be taken offline for an upgrade, you may need a different approach (like a hybrid overlay). Boards should expect a clear output here: a prioritized roadmap that says "System X (high risk) – PQC pilot in next 12 months; System Y (medium risk) – upgrade in 2-3 years," and so forth.

3. **Plan Pilot Projects** – Before diving into organization-wide changes, run pilots of quantum-safe solutions in controlled environments. Pilots serve as **proof-of-concept and learning opportunities**. For example, you might choose a non-customer-facing internal application and implement a PQC-enabled version of its encryption module, or set up a test network that uses a hybrid VPN with PQC. The goal is to observe **real-world impacts**: How did performance change? Did any compatibility issues arise? What new skills or tools did the team need? Use these pilots to refine your transition plan. It's wise to involve a variety of teams – security architects, network engineers, application developers, compliance officers – to get a 360-degree view of the implications. Successful pilots build confidence (for both technical staff and executives) that the new technology will work. They also help **develop internal expertise**. By the time you scale up, you want your team to have hands-on experience. From the board's perspective, ask management for pilot outcomes: *"What did we learn from our PQC pilot? Have we documented best practices and pitfalls?"* Treat pilots as a required milestone before large investments.

4. **Execute a Phased Rollout**: With lessons learned and a plan

in hand, begin the broader implementation **iteratively**. A big-bang crypto swap is risky; instead, update systems in phases aligned to the prioritization from step 2. This might mean first upgrading all public-facing channels (like your websites' TLS encryption and customer portals) to quantum-safe modes, then core internal systems, and so on. During rollout, **maintain dual support** where possible: e.g., enable both classical and post-quantum algorithms and gradually shift the default to PQC once confidence is high. Key management infrastructure (Public Key Infrastructure, key management services) will likely need updates early in this phase, since they underpin everything else. Also, plan for **contingencies**: for instance, if a chosen PQC algorithm later shows a weakness or performance issue, have an alternative ready (this is where crypto-agility design pays off). Throughout the rollout, it's critical to monitor for any issues in security **and** availability; any sign of trouble might mean pausing to adjust. Regular status reporting to the board is advisable here, to keep leadership apprised of progress and any needed course-corrections.

5. **Governance and Continuous Adaptation** – Achieving a quantum-safe state isn't a one-and-done project; it becomes an ongoing aspect of your security governance. Boards and executive teams should institute **metrics and oversight** for cryptographic health. This can include KPIs like "percentage of systems quantum-safe," "number of legacy crypto instances remaining," or how quickly new cryptographic updates are applied. A governance framework might involve a **cryptography steering committee** or extending the remit of the risk committee to cover crypto-agility. Additionally, **update your policies**: for example, embed quantum safety into your software development lifecycle (require new systems to use approved PQC algorithms), and into vendor procurement (more on that soon). Ensure there is an owner (e.g. a **head of cryptographic services** or your CISO) accountable for keeping the organization's encryption current. Regular audits or reviews, possibly with third-party experts, can validate that your crypto deployment is sound and that no shadow IT or forgotten systems are undermining your safety. Remember, threats will continue to evolve; perhaps a decade from now we'll be discussing *post*-PQC if new quantum

capabilities emerge. By building a culture of crypto-agility and strong governance now, you prepare the organization to handle whatever comes after PQC as well.

Throughout this roadmap, maintain clear **communication and training**. Front-line tech teams need education on PQC tools and techniques (as new protocols or libraries roll out), and non-technical stakeholders should be kept in the loop on why these steps are important. Employees are more likely to support crypto changes (which can sometimes cause short-term inconvenience) if they understand the *why*: that it's protecting the company's future and customers. In essence, treat quantum resilience as you would a major organizational change initiative: strong executive sponsorship, good project management, cross-functional collaboration, and transparency.

Internal Alignment: Breaking Silos for Crypto Migration

Implementing the quantum-safe roadmap is not solely the CISO or IT department's job – it requires **broad internal alignment**. Organizations that succeed in such transitions do so by uniting stakeholders from across the enterprise. Here's who needs to be on board and how to get them aligned:

- **Executive Leadership & Board:** Set the tone at the top. Boards should explicitly include quantum risk in their risk oversight, and executives should make PQC preparedness a stated priority (e.g., in strategy memos or OKRs). This top-down mandate is critical; otherwise, middle management may defer action in favor of more immediate fires. Leadership's role is to **champion resources** (budget, personnel) for the PQC effort and to hold teams accountable to timelines. It's wise to assign an executive sponsor – say, the CIO or CISO – whose performance objectives include delivering the quantum-safe program.
- **Cryptography Center of Excellence / Task Force:** Many leading companies are creating a **crypto-agility or quantum-readiness task force** to drive this effort. This is essentially a cross-functional working group that brings together experts from security, IT infrastructure, application development, risk management, and even legal/compliance. The task force acts as a coordination hub – developing the strategy, tracking progress, and troubleshooting obstacles. Crucially, it should report to senior leadership (e.g., give updates to the risk

committee or CIO monthly) to ensure visibility. If your organization is large or in a regulated industry, consider formalizing this as a "Cryptographic Center of Excellence," as has been done in some financial institutions. This sends a message that crypto agility is here to stay as a core competency, not a one-off project.

- **Security and IT Teams:** These teams are on the front lines of implementation. To align them, emphasize that quantum migration isn't just extra work – it's an opportunity to **modernize and fortify** the entire security architecture. Often, crypto updates coincide with good housekeeping: replacing hard-coded secrets, improving key management systems, retiring obsolete protocols (like old versions of TLS) and so on. Encourage a mindset of *"fix the roof while the sun is shining."* That means addressing current crypto weaknesses (e.g. any use of deprecated algorithms like SHA-1) under the umbrella of the quantum program. It's also vital to **train** these teams. Provide learning sessions on how the new algorithms work, how to use libraries that implement them, and pitfalls to avoid. Build internal knowledge so that your staff feels confident with PQC by the time it's production-critical.
- **Application Developers & Product Teams:** If your company builds software products or customer-facing applications, those teams must integrate quantum-safe practices into their roadmaps. This might mean updating SDKs and libraries to ones that support PQC, refactoring parts of code that assume fixed key sizes or algorithm types, and generally ensuring new development is "quantum-ready." Product managers should factor in that customers (especially in B2B or government segments) will soon demand quantum-safe security as a requirement. Being able to market your software or service as **quantum-resistant** could become a competitive advantage in the next few years. Internally, align these teams by highlighting that proactively adopting PQC can prevent future scramble (or fire drills) to patch security. It's much harder to retrofit quantum-safety into a product at the last minute than to design with flexibility from the start.
- **Procurement and Vendor Management:** Your procurement policies should be updated to drive internal alignment on crypto resilience. Put simply, whenever you buy

a product or service that involves cryptography, you should be asking vendors about their **PQC roadmap**. Procurement can include language in RFPs and contracts requiring suppliers to support approved post-quantum algorithms by a certain date, or to attest to their crypto-agility. Internally, this aligns the business side (which often owns vendor relationships) with security goals. The message: we won't invest in technology that could become a security liability in a few years. By pushing this criterion, you also raise awareness – business and procurement managers start to understand quantum risk because it's now part of the checklist when evaluating solutions. Many organizations learned this with cloud security and privacy requirements; now we must do the same with quantum safety.

- **Legal, Compliance & Risk Officers:** These stakeholders ensure that quantum resilience efforts align with regulatory requirements and do not introduce undue risk. They should be consulted early to identify any compliance constraints (for instance, certain regulations might mandate FIPS-validated cryptography – you'll need to ensure chosen PQC algorithms meet those). Also, the legal team can help update any customer-facing commitments or disclosures about data security to include quantum-safe terminology when appropriate. Aligning here also means planning for how to **audit and document** your crypto transition – proving to auditors, regulators, or partners that you took prudent steps ahead of the threat. In industries like finance or healthcare, showing quantum preparedness could soon be seen as part of fiduciary duty or standard due diligence. Getting compliance folks on board early will make external assurance smoother later.

The common thread in internal alignment is **communication**. Every relevant team should know the *why*, *what*, and *when* of your quantum-safe program. Consider an internal newsletter or dashboard for the project. Celebrate milestones (like "we just completed our first PQC-encrypted transaction" or "100% of external websites now using hybrid TLS"). This keeps momentum and buy-in. And from the top, reinforce that **being safe is part of being a trusted, modern enterprise** – not just an IT chore. That cultural framing will help break silos; people will see this as a company-wide resilience initiative.

Securing the Ecosystem: Managing External Dependencies

No organization is an island when it comes to cryptography. You rely on a multitude of external partners and providers – from the software vendors whose products you use, to cloud and SaaS services, to hardware manufacturers and beyond. Your quantum resilience is only as strong as the weakest link in this extended ecosystem. Executives and boards must therefore **drive quantum-safe readiness outward** into their supply chain and partnerships.

Start by taking the vendor list from your cryptographic inventory (Step 1 of the roadmap) and **engaging each critical supplier in a conversation about PQC**. Ask them: *What is your plan to upgrade your product/service to support post-quantum encryption and signatures?* The answers should be factored into your risk assessment. If a key vendor has no clear plan, that poses a risk to you – you may need to apply pressure or consider alternatives down the line. Some vendors are ahead of the curve (for example, several cloud providers and network equipment makers are already testing PQC in their offerings), but many are not. By raising the question, you also signal that this is important to customers – which in turn nudges vendors to prioritize it. Remember, vendors have their own competing priorities; vocal customer demand for quantum-safe features will elevate its importance.

For strategic vendors, consider formal steps: **require PQC commitments in contracts.** Just as contracts often include security requirements (like compliance with certain standards), include a clause about maintaining crypto-agility or supporting new encryption standards. One industry advisory bluntly states: *"Organizations cannot secure their systems without their vendors first providing the necessary cryptographic libraries, certificates, and protocols"*. In other words, if a critical component (say a database system or an ERP platform) doesn't offer PQC options, it doesn't matter that you want to be quantum-safe – you'll be stuck. Thus, make it a **partnership effort**: maybe coordinate joint testing with a vendor's beta version of a PQC feature, or co-host a workshop with key suppliers on quantum readiness. The more you collaborate, the smoother your own transition. Leverage industry groups as well – for instance, if you are in a sector consortium (like FS-ISAC in finance or NH-ISAC in healthcare), work together to put collective pressure on common suppliers to expedite quantum-safe roadmaps.

Cloud providers warrant special mention. Many enterprises have consolidated their IT on a handful of cloud and SaaS platforms. Check those providers' public statements or documentation: some have published their PQC migration plans or even started offering hybrid TLS connections. If a major cloud you use hasn't articulated a plan, ask them. As a customer, you have influence, especially if you're an enterprise client. **Don't be shy about making quantum safety a key performance indicator for your vendors.** Some forward-looking organizations are already doing this – issuing questionnaires to vendors about crypto-agility, just like they do for data privacy or uptime SLAs.

Executive Signal: Make quantum safety a supplier requirement – if your partners aren't quantum-ready, neither are you.

Another external dependency is the **public-key infrastructure (PKI)** that underpins trust on the internet and between organizations. Certificate Authorities (CAs) will need to offer quantum-safe certificates; standards for those are being worked out (e.g., how to embed PQC public keys into X.509 certificates, how to do cross-signing with classical CAs). Ensure your teams follow these developments. You might not run a CA, but you certainly buy certificates. In the near future, you'll want the option to get, for example, a hybrid certificate (one that can be validated by both classical and PQC methods) for your public website, or to secure inter-company APIs. Some CAs have begun trials of such certificates. Keeping an eye on this and being ready to adopt them early – even if just in a test environment – will keep you ahead of the curve. It also again signals to the market that customers care about this.

Finally, consider collaborating beyond just vendor-client relationships. **Industry alliances and government initiatives** around quantum-safe cryptography are proliferating. By participating, your organization can help shape solutions and stay informed. Whether it's a tech consortium creating interoperability profiles for PQC, or a government task force issuing migration guidelines, these forums ensure you aren't solving problems in isolation. They can also amplify your voice in pushing for things like common standards (which can ease your vendor management burden). Many sectors have started quantum working groups precisely because of the recognition that one company's lapse can affect many (think of a certificate authority breach or a compromised software update mechanism – these can have systemic effects). So being a team player externally is part of being safe, too.

In summary, extend your resilience strategy to every entity that handles your data or connects to your systems. As you harden your own crypto, make sure the ecosystem is coming along. This external engagement is as much a leadership task as an operational one – it requires executives to use their influence in the industry, and perhaps to invest in joint solutions where needed. The payoff is a **collectively stronger defense** and fewer weak links that could be exploited via the backdoor.

Risk Reduction and Reputational Value: Security as a Trust Signal

Investing in quantum resilience isn't just about avoiding future hacks – it's also about **protecting your brand and capitalizing on trust**. In the digital era, security underpins trust, and trust underpins business. Customers, investors, and regulators are increasingly savvy about cyber risks, and quantum is entering that conversation. Demonstrating that you are "quantum-safe" can become a **differentiator and a reassurance**. In fact, encryption itself is a visible trust signal (the little lock icon in your browser's address bar, for example); moving to quantum-safe encryption is the next evolution of that signal. As one industry analysis noted, *"Beyond technical compliance, encryption is a trust signal. Falling behind on the transition to post-quantum-safe encryption risks damaging a company's reputation with customers, investors, and partners who are increasingly alert to cybersecurity issues. Inaction may be interpreted as complacency or even negligence."*. In plain terms: if you delay and eventually there's a quantum-related breach or scramble, stakeholders might say **"Why didn't you act when you knew the risk?"** On the flip side, being proactive lets you tell a positive story about resilience and innovation.

Consider the **risk reduction angle**: By migrating to stronger cryptography, you are lowering the probability of catastrophic data breaches in the future – specifically the kind that could occur silently if adversaries decrypt old data or forge credentials once quantum computing hits a threshold. This directly protects against financial losses (from breach response, fines, lawsuits) and also against the **secondary impacts** like loss of competitive information or IP. Some sensitive data, if leaked, could erode your competitive advantage or public trust permanently. Quantum-proofing such data is like buying a very advanced insurance policy: you hope the scenario never happens, but if it does, you've drastically mitigated the impact.

Speaking of insurance, the cyber insurance industry is starting to pay attention to quantum risk. Insurers are in the business of pricing risk, and they foresee that clients stuck on outdated encryption may be a higher claim payout waiting to happen. Forward-thinking insurers may begin to **incentivize quantum readiness** – perhaps by offering lower premiums or enhanced coverage to companies that can demonstrate crypto-agility and PQC adoption. By being an early mover, you position your organization to benefit from such incentives. Even if direct premium discounts are not yet formalized, you'll stand out in insurance questionnaires (which increasingly ask about security practices). A company that can say "Yes, we have an active program to upgrade our cryptography and we're tracking to be quantum-safe by X date" will be viewed more favorably than one that says "No action yet, just watching." It's analogous to how insurers consider fire safety measures in a building – better to have sprinklers installed before the fire.

Now think of the **customer and market trust** dimension. If you operate in a consumer space, your marketing team might not run ads about post-quantum cryptography – it's too esoteric for the average buyer. But large enterprise customers and partners *will care.* If you're a bank, for example, corporate clients and institutional investors will be keen to know that their data (which might have long-term sensitivity) is safe from emerging threats. If you're a cloud provider or software vendor, you can bet that within a couple of years RFPs will include questions about quantum-safe security. Being able to check that box confidently (or better, to help educate your clients on it) gives you an edge. It says *our product is built for the future.* Even internally, imagine the morale and culture boost for your security and IT teams in being regarded as industry leaders on this front. It helps attract talent (top security professionals want to work where security is taken seriously and innovatively).

Importantly, **being safe is part of being ahead**. The name of this book is *Quantum Advantage: Be Ahead. Be Safe. Win.* – these themes reinforce one another. By taking the initiative on quantum resilience ("Be Safe"), you inherently are acting ahead of many competitors, which could translate into business advantage ("Win"). You're not scrambling at the last minute, so you can focus on seizing opportunities that quantum tech will create (the positive side of the quantum revolution, which we'll explore in later chapters). You're also avoiding the scenario of being caught in a reactive, costly rush – the scenario some organizations experienced during past incidents like the Y2K bug or after sudden breaches where everyone had to patch overnight. Instead, you're steadily executing a multi-year plan under your control, which is typically far cheaper and smoother than emergency fixes.

Analysts have noted that **delaying the PQC transition will only increase costs and risks** – a rushed migration in crisis mode could lead to mistakes and higher expenses. Starting now is a **strategic investment** in stability and future savings.

Finally, the reputational value extends to the public and regulatory domain. There may come a time (imagine a news headline like "Quantum computer built that can crack encryption") when consumers broadly become aware and concerned. Companies that can swiftly assure, "We have already upgraded our security to quantum-resistant standards," will weather that moment far better than those who cannot. Regulators, too, might start questioning firms on their preparedness (much like they do for disaster recovery or climate risks in some sectors). Being on the front foot means you have answers ready and can even help shape sensible regulations rather than be caught off guard by them.

To sum up: **being safe is smart business**. It reduces tangible risk, protects your brand's trust equity, meets emerging compliance expectations, and can even yield financial benefits (through insurance or customer preference). In the eyes of your stakeholders, a commitment to quantum resilience signals that your organization is not just reacting to threats, but leading on security and reliability. In an era where digital trust is paramount, that's a profound competitive advantage.

As you champion the quantum resilience playbook, remember that this is a journey of leadership. It's about steering your organization through a coming era of change – calmly, confidently, and proactively. "Be Safe" is not a static end state but a continuous posture of vigilance and agility. By instilling crypto-agility, embracing new standards, leveraging hybrids, executing a solid roadmap, aligning your entire organization, and extending security to every partner, you are building a **digital fortress ready for the quantum age**. In the next chapter, we'll shift focus from defense to offense – exploring how to *Win* by leveraging quantum opportunities. But no offensive strategy works without a solid defense. With the resilience steps from this playbook, you are ensuring that your digital trust foundation remains unshakeable, no matter what technological disruptions come your way. In doing so, you uphold the trust of those who depend on you – customers, employees, partners, and society – proving that proactive security is part and parcel of visionary leadership.

6 BE AHEAD: STRATEGIC ADVANTAGE THROUGH EARLY ACTION

In technology, timing is everything. The organizations that act early to embrace new paradigms often reap outsized rewards in competitiveness and resilience. We saw this with the shift to cloud computing – firms that migrated early gained productivity boosts, revenue growth, and a foundation for adopting future innovations. The same pattern emerged with artificial intelligence: recent research shows *early adopters of generative AI are already "reaping significant rewards, from increased revenue, to better customer service, to improved productivity"*. Whether it was mobile-first services or the e-commerce revolution, those who moved first achieved differentiation while laggards fell behind. **Quantum** is no different. Though still emerging, it is poised to **transform computing and security** – and **first movers will gain a strategic edge**. Early action on quantum readiness isn't just about hedging risk; it's about being the organization that shapes the future and leads its industry into the next era.

The Early-Mover Advantage in Quantum

History suggests that **early movers win**. In quantum technology, this principle will be even more pronounced. Executives should recall how early cloud adopters became more agile and innovation-ready than peers, or how pioneers in AI secured patents and talent that latecomers couldn't match. Likewise, *first movers in quantum stand to secure unique benefits*: they can patent new quantum algorithms and applications before others, establish partnerships and intellectual property that lock in long-term advantage, and build **capabilities years ahead** of competitors. A recent industry analysis put it plainly: *"quantum readiness comes with a first-mover advantage and the power to shape the direction of the technology."* Organizations that start exploring quantum use cases today – even if the technology is nascent – will be positioned to **harness its full power as it scales**. By the time quantum computing reaches commercial viability, these trailblazers will have operational experience, optimized workflows, and a pipeline of quantum-skilled talent, whereas late adopters will be scrambling to catch up. In short, **being ahead means building durable competitive differentiation now**.

Critically, the field is still wide open. Few enterprises have taken substantive action on quantum yet. In one survey, only **12% of business leaders felt their organizations were prepared to even assess quantum opportunities**. IBM's global "Quantum-Safe Readiness Index" similarly found that most companies score very low (an average of 21 on a 100-point scale), with only a top 10% of "Quantum Champions" making significant early progress. This means an executive bold enough to act in 2025 can **join a very select leadership group**. The rarity of preparedness turns early adoption into a powerful story to tell stakeholders. It signals foresight and tech leadership. Conversely, waiting carries strategic peril: as quantum capabilities mature, the gap between early adopters and laggards will become an innovation chasm that is extremely difficult to bridge. The message is clear – **act now, lead now**.

Talent: Attracting Quantum-Ready Minds

One of the most immediate advantages of early action is access to **scarce talent**. Quantum computing and post-quantum cryptography demand highly specialized skills – from quantum scientists and cryptographers to engineers fluent in new algorithms. Such experts are in *incredibly short supply*, with only a handful of academic groups globally producing qualified candidates. Organizations that move early on quantum initiatives send a powerful signal to this talent pool. By launching quantum projects, partnering with universities, or funding research, you *become a magnet for quantum-savvy professionals*. Early movers can build relationships with top minds while competitors are still unaware of the need. As one talent recruiter observed, the race for quantum experts is accelerating: *"With more investment comes more interest, with more interest comes more competition… as the market becomes more competitive,* ***time is of the essence*** *on moving on candidates."* Simply put, if you don't snap up quantum-literate talent now, your rivals or other industries will.

Beyond hiring, taking early action helps **cultivate your existing workforce.** It creates internal opportunities for engineers and analysts to grow into quantum and cryptography roles, improving retention of ambitious employees. Leading organizations focused on quantum readiness are already fostering a *"thriving talent ecosystem"* around these emerging skills. They sponsor employee training in quantum concepts, encourage cross-functional teams to experiment with quantum algorithms, and establish mentorship networks with academia. This not only addresses the talent shortage proactively, but also future-proofs the organization's skill base. Remember, every major tech shift – from big data to AI – saw a war for talent. Quantum is no exception. By acting early, **you win the talent war before others even show up**. In the boardroom, the ability to say *"we have one of the leading quantum security teams in the industry"* is a strategic bragging right that few others can claim.

Trust and Resilience: Earning Stakeholder Confidence

Another strategic benefit of early quantum readiness is the **trust and confidence it inspires** among key stakeholders. Customers, business partners, auditors, and regulators are all increasingly aware of the coming quantum disruption – especially the threat it poses to cybersecurity. They know that data encrypted with today's standards could be vulnerable in the near future. An organization that can credibly say *"we are already quantum-safe"* immediately differentiates itself as a forward-looking, trustworthy partner. Early adoption of quantum-safe practices **maintains customer trust** by assuring them that their sensitive information will remain protected even as technology evolves. In sectors like financial services or healthcare where privacy is paramount, this assurance isn't just technical – it's a **brand promise**. It tells the market that your firm is not asleep at the wheel but is instead anticipating and neutralizing tomorrow's threats.

Proactive quantum readiness also earns **regulatory goodwill**. Around the world, regulators are beginning to urge companies to prepare for quantum risks in order to uphold security and stability. For example, the European Union's Digital Operational Resilience Act (DORA) explicitly calls for financial entities to stay *"abreast of developments in cryptanalysis"* and adopt a **"flexible approach"** to mitigate new cryptographic threats like quantum. Supervisors don't want to see another industry scrambling at the last minute (as happened with past security crises); they prefer to see calm, preventive action. By moving early, you position your organization as a *responsible leader* that regulators can rely on, rather than a laggard that might need heavy-handed compliance intervention later. Early movers may find that regulators give them a lighter touch, involve them in shaping guidelines, or publicly commend their leadership. In practice, **being ahead on quantum resilience now could translate into an easier time with auditors and examiners later**. It's an investment in your license to operate. In short, showing quantum readiness isn't just about protecting data – it's about protecting and enhancing the **trust** others place in your enterprise.

Innovation Posture: Ready to Leverage Quantum Opportunities

"Be Ahead" is not only about defense; it's also about **offense** – positioning your organization to seize new opportunities that quantum technologies will unlock. Early action on quantum readiness naturally cultivates a forward-looking **innovation posture**. By engaging with quantum-safe cryptography, quantum algorithms, or even modest quantum computing pilots today, your company builds familiarity and technical fluency in an area that will revolutionize products and processes tomorrow. Organizations that invest now will be **first in line to explore quantum applications** as they become viable. For instance, once breakthrough quantum hardware arrives, an early-moving bank that has already experimented with quantum optimization algorithms can quickly apply them to portfolio management or risk analysis ahead of competitors. A logistics firm that has dabbled in quantum routing models will implement next-gen optimization faster than peers who have to start from scratch in a few years.

Being ahead on quantum also forces an organization to develop valuable **adjacent capabilities**. Preparing for quantum-safe encryption requires improving your cryptographic agility, asset management, and IT architecture. These improvements have spillover benefits – simplifying your infrastructure and making you more adaptable to any emerging tech. The same teams and processes that manage a crypto-agility project today could manage a quantum-computing integration project tomorrow. In essence, you are *building the muscles* for innovation. Leaders at Microsoft Azure's Quantum program note that companies planning a comprehensive quantum-ready strategy now are *"creating durable, competitive differentiation"* and **positioning themselves to harness the full power of quantum as it scales**. They are learning how to marry quantum capabilities with classical IT, how to identify high-value use cases, and how to cultivate internal champions for cutting-edge tech. These are exactly the kind of organizations that will not only defend their market position, but **expand it**, when quantum breakthroughs arrive.

Consider also the cultural signal. Embracing quantum readiness nurtures a culture of **curiosity and experimentation**. It tells your workforce that innovation is part of the company's DNA and that leadership is willing to invest in transformative technology – even if immediate ROI is not guaranteed. This mindset pays dividends beyond quantum: it keeps your company adaptable and alert to other disruptive innovations (whether in AI, biotech, or beyond). In summary, early quantum action isn't a sunk cost; it's an **innovation catalyst**. It ensures you won't be the executive waking up to a quantum-driven industry disruption in 2028 wondering, "why weren't we ready?" Instead, you'll be the one orchestrating that disruption to your advantage.

Shaping Standards and Protocols: A Seat at the Table

When a technological frontier is in flux, early adopters get something priceless: **influence**. By acting now on quantum initiatives, your organization can help shape the emerging standards, protocols, and ecosystems that will govern the quantum era. Early movers often earn a *seat at the table* in industry consortia, standards bodies, and pilot programs. We've seen this pattern with previous tech waves – the companies that joined early working groups for internet protocols or mobile network standards ended up steering those conversations to their benefit. In quantum security, for example, businesses that start implementing post-quantum cryptography (PQC) can provide real-world feedback to standards organizations like NIST and ETSI, potentially influencing how algorithms are fine-tuned or which transition approaches are favored. Governments and industry groups are actively seeking **pilot participants** to test quantum-safe methods and quantum communications; if you volunteer early, **you help shape the rules** (and maybe even secure favorable terms or first access to breakthroughs).

The advantage extends to policy influence. Regulators crafting frameworks for a quantum world will lean on the experience of first movers. If your firm is known as a quantum-ready leader, don't be surprised if you're invited to advisory committees or cited in policy discussions. This gives you a chance to advocate for pragmatic transition timelines, reasonable compliance burdens, and incentives for innovators. In essence, you can help ensure the playing field that's coming is one you're already prepared for. As academic observers have noted, being quantum-ready early confers *"the power to shape the direction of the technology"*. This might mean defining best practices that others follow, patenting foundational techniques, or leading consortiums that set technical guidelines. Such influence can yield long-term strategic control – think of it as writing the playbook for a game that you intend to win. For executives, this isn't just a technical detail; it's **strategy at the highest level**. You are ensuring that as quantum tech matures, your company's perspective and interests are woven into the very fabric of how the industry operates.

Regulatory Goodwill: Proactive Today, Preferred Tomorrow

Every executive knows that regulatory landscapes can either constrain or enable business strategy. Early quantum readiness can tilt the balance in your favor by earning **regulatory goodwill**. We touched on trust and lighter oversight, but it goes further – it's about being seen by regulators as part of the solution, not part of the problem. Governments are increasingly aware of quantum risks (and opportunities) and are starting to push for action. In the United States, for instance, federal directives now require agencies to inventory their cryptographic systems and prepare migration plans for PQC. In financial services, as noted earlier, the EU and other jurisdictions are incorporating quantum-safe expectations into resilience regulations. An organization ahead of these mandates not only avoids the scramble of compliance when new rules hit, but can actually *shape those rules* through early example.

Regulatory goodwill shows up in various ways. It could mean faster approvals or certifications because your security posture is demonstrably advanced. It could mean more productive, less adversarial relationships with examiners who see that your team "gets it" and is forward-thinking. It might even mean **competitive advantages codified into contracts**, as customers in regulated sectors (like government procurement or banking partnerships) prefer vendors who meet upcoming standards. For example, if a major bank must ensure all vendors will be quantum-safe by 2030, being able to say *"we're already there"* puts you at the front of the line for those deals. Early action can also lead to **policy influence** as mentioned – regulators often pilot new guidelines with industry leaders to work out kinks. If you're proactive, you could help design frameworks that suit your capabilities (and raise the bar for competitors). In sum, **early compliance is cheaper and carries side benefits**. You invest now in upgrading crypto and processes, and in return you potentially get years of smoother sailing on the regulatory front. It's a classic case of an *ounce of prevention* being worth a pound of cure, especially when that prevention builds goodwill with those holding the whistle.

Momentum Across Industries (Finance, Pharma, Telecom, and Tech)

Lest one think quantum readiness is a concern in isolation, it's worth noting that **multiple industries are already stirring**. Forward-looking leaders across sectors are starting to make their moves – providing real examples to learn from and a clear signal that early action is feasible (and happening).

- **Financial Services:** The finance industry, highly sensitive to security and speed, is among the first to act. Major banks and exchanges are conducting trials of quantum-safe security today. For instance, HSBC (one of the world's largest banks) has partnered with regulators in Singapore on a pilot of quantum key distribution for secure banking communication. In 2022, BT launched one of the world's first quantum-secured networks in London, and early customers included HSBC itself and EY, validating demand in finance and professional services. Banks are also on the offensive with quantum computing experiments: JPMorgan and Goldman Sachs have publicly explored quantum algorithms for portfolio optimization and trading risk. The message is clear – in finance, being ahead on quantum is increasingly seen as essential to protect assets and **out-compute** the competition.

- **Pharmaceutical & Healthcare:** Pharma companies are eyeing quantum not just for security but for R&D breakthroughs. Quantum computing's ability to model molecular interactions promises to accelerate drug discovery dramatically. That's why giants like Merck, Johnson & Johnson, Roche, and Amgen have already **filed patents related to quantum computing applications**. In the last couple of years, we've seen high-profile partnerships: Moderna joined with IBM to explore quantum technologies for mRNA medicine, and AstraZeneca and Sanofi teamed with SandboxAQ (an Alphabet spin-off) to begin integrating quantum approaches. These early adopters aim to shorten research cycles and gain an edge in discovering new treatments. Healthcare providers, too, are interested in quantum-safe data protection for patient records. The sector may be in early innings, but clearly some of the biggest players believe early quantum investment will lead to **life-saving and market-winning innovations**.

- **Telecommunications:** Telcos form the backbone of digital infrastructure, so it's no surprise they're among early quantum movers, especially in communications security. Telecom operators in advanced markets are rolling out **quantum-safe networks** as premium offerings. In South Korea, SK Telecom started quantum initiatives as far back as 2018, and by 2025 Korea Telecom launched a commercial quantum-secured VPN over its 5G network . In Europe, British Telecom's 2022 QKD service in London was a trailblazer, and in 2025 Orange (France) unveiled "Orange Quantum Defender," a quantum-secured metro network in Paris combining QKD and post-quantum cryptography . They've already signed up financial institutions and are in talks with defense and aerospace clients for this service. Meanwhile, operators like Telefónica, Singtel, Telus, and Turkcell are all publicly experimenting with quantum communications. The telecom industry's play is clear: **offer quantum-grade security** as a differentiator, and learn now so that when a "quantum internet" eventually emerges, they will operate it. Early action here is both a revenue opportunity and a defensive move to secure the networks against future attacks.
- **Technology & Cloud Providers:** The tech sector itself is both driving and adopting quantum advances. All major cloud platforms – IBM, Google, Microsoft, Amazon – now offer some form of "Quantum-as-a-Service," allowing companies to experiment with quantum computing in the cloud. These providers are in a race to enable enterprise quantum readiness. Tech companies are also leading on integrating post-quantum cryptography into existing products. Google, for example, implemented a hybrid post-quantum TLS algorithm in Chrome 116 in 2023 to start protecting web traffic. Companies like Cloudflare and IBM have similarly tested quantum-safe encryption in their services. Startups such as QuSecure demonstrated a **quantum-resilient satellite link** using SpaceX's Starlink satellites in 2023 – a hint of quantum-safe

communication for the future Internet. What this means for an executive is that the tools for quantum readiness are rapidly becoming available off-the-shelf. The broader tech ecosystem is moving so that early adopters in any industry can plug into quantum solutions (be it through cloud APIs, new security libraries, or partnerships). In short, *you won't be alone*: from Wall Street to pharma labs to telecom hubs, **the early movers club is growing**. The question is, will you join it now or watch from behind as your peers forge ahead?

Signals from the Top: How Leaders Drive the 'Be Ahead' Mindset

Achieving a strategic advantage through early action requires **strong signals from the executive level**. It's not enough to have an innovation lab dabbling in quantum; leadership must actively champion the "be ahead" agenda so that it permeates the organization's strategy and culture. For boards and C-suites, this means using your sponsorship and oversight to ensure quantum readiness gets the attention it warrants. Here are several concrete ways to lead from the front:

1. **Invest and Sponsor Boldly:** Treat quantum readiness as a priority initiative, not a research curiosity. Allocate dedicated budget for quantum-safe cryptography upgrades and quantum pilot projects. For example, sponsor a company-wide **cryptographic inventory and upgrade program** so that every critical system will be quantum-safe by a target date. Fund small teams to run proofs-of-concept on quantum computing platforms (available via cloud) for high-value problems. Executive sponsorship shows the organization that this is a *mission*, not a hobby.

2. **Appoint Accountable Leaders:** Establish clear ownership of the quantum readiness agenda. Many firms are now creating roles like Head of Quantum Strategy or adding quantum responsibilities to the CIO/CISO's mandate. Ensure someone at the senior level wakes up every day thinking about quantum threats and opportunities. As surveys indicate, companies that have already appointed quantum leaders and pilot teams are far more likely to make progress. By appointing an executive or task force and empowering them, you signal that **quantum is "owned" at the top**.

3. **Integrate Quantum into Strategy and Risk Planning:** Insist that quantum considerations be embedded in enterprise risk management, IT roadmaps, and innovation strategy. For instance, when reviewing the IT strategy, ask: *"Have we planned for post-quantum encryption in our systems upgrade cycle?"* In product discussions, inquire: *"Could quantum computing disrupt this business model, and how will we capitalize on it if so?"* By asking these questions regularly, you normalize the topic. Make "quantum readiness" a line item in strategic planning templates and board risk reports, right alongside cybersecurity and AI. If it's part of the planning process, it won't slip through cracks.

4. **Champion Skilling and Culture:** Encourage HR and learning teams to launch **quantum literacy and cryptography training** for technical staff (and even for interested managers and board members). The goal is to demystify quantum and build a pipeline of internal talent. Celebrate early quantum projects and the teams behind them. Consider rotations or fellowships where top IT talent can spend a few months with a quantum research group or in a security lab exploring PQC. Culturally, reward proactive behavior – if an engineer finds a creative way to implement a quantum-safe solution, hold it up as an example of the "ahead" mindset. Build excitement that *"we are pioneers"*. This keeps morale high and fear low; employees won't dread quantum change, they'll drive it.

5. **Engage Externally and Set the Narrative:** As a leader, use your platform to position your organization as quantum-forward. This could mean speaking at industry events about your quantum readiness efforts, joining cross-industry consortia, or even collaborating with competitors on pre-competitive standards (security is a shared interest). Challenge your enterprise: *"Will we be the first to answer a customer's quantum-security questionnaire or RFP?"* In many industries, large clients or government agencies will soon start asking suppliers how they are mitigating quantum risk. Make sure your company can answer confidently – ideally **being the first in your sector to say "we're quantum-ready"** when others stammer. This not only wins business but also reinforces your brand as an innovator and trusted partner.

Collectively, these actions embed the "ahead" mindset into the company's DNA. The common thread is **intentional leadership**: by visibly driving the quantum agenda, executives remove the ambiguity and inertia that often plague new initiatives. Instead of waiting for a bottom-up project to gain traction over years, you create a top-down mandate that *accelerates readiness now*. As Google's research on AI adoption found, *91% of organizations with strong C-level support saw significant performance gains* – the same principle applies here. Your sponsorship converts quantum from a distant issue into a current strategic priority. And importantly, it gives your people permission to think big and act early. The tone from the top can inspire an entire organization to **be proactive, be bold, and be ahead**.

Conclusion: Ahead Now, Ready to Win Next

Taking early action on quantum readiness is a definitive leadership move. It's about seizing the initiative – **being the disruptor, not the disrupted**. By moving early, you fortify your enterprise against tomorrow's threats *and* poise it to exploit tomorrow's breakthroughs. You attract the best talent, assure your customers and regulators, and set your organization on a path of continual innovation. This is how companies turn technological change into strategic advantage: not by reacting at the last minute, but by anticipating and preparing well in advance.

As you drive your organization to be ahead, keep the endgame in mind. Quantum readiness is not just a defensive play; it's the foundation for *offensive wins* in the near future. The next chapter will delve into exactly that – how being secure and prepared opens the door to **transformative opportunities and new value creation**. After all, the book's message is *"Be Ahead. Be Safe. Win."* You've seen how being ahead grants safety and strength. In **Chapter 7: Win**, we will explore how those who lead early can translate their quantum advantage into market leadership, new products, and growth. Winning isn't just about surviving the quantum upheaval; it's about *capitalizing on it*. By acting today, you aren't merely avoiding a downside – you are investing in an upside that others can barely imagine.

In closing, remember that **foresight is a hallmark of great leadership**. Just as the best CEOs of the past decade championed digital transformation before it was a buzzword, today's leaders will champion quantum readiness before it becomes an urgent mandate. They will be the ones whose companies are not only secure when the quantum reckoning comes, but are already using quantum technology to outperform competitors. Early action is an act of confidence and vision. So be bold, take those first steps, and carry your organization forward. **Be ahead** – and set the stage to be safe and to win in the quantum era. Your strategic advantage tomorrow begins with your early action today.

7 WIN: TURNING QUANTUM READINESS INTO BUSINESS ADVANTAGE

Quantum readiness isn't just about defense – it's a strategic asset. In this chapter, we explore how being **quantum-ready** transforms from a cost center into a competitive differentiator that drives revenue, trust, and growth. By investing early in quantum-safe security and emerging quantum technologies, organizations can **monetize their readiness**, signal strength to the market, unlock new opportunities, and build a compelling business case for long-term advantage. The message is clear: *Be Ahead. Be Safe. Win.* – not only to survive the quantum era, but to thrive in it.

Monetizing Quantum Readiness

1. Premium Security as a Differentiator: Companies that proactively adopt quantum-safe security can command a premium in the market. Much like early adopters of EMV chip cards touted safer payments, today's early movers can market **"quantum-safe"** products and services as superior offerings. For example, a financial institution or SaaS provider that enables **quantum-safe encryption** for client data can advertise this as a **premium feature**, reassuring customers that their information is protected against future quantum threats. This differentiation not only justifies premium pricing or contract wins, but also sets a high bar that competitors must race to meet.

2. New Revenue Streams and Services: Quantum readiness can itself be productized. Forward-thinking firms are creating **quantum-secure product lines** and services – from quantum-safe cloud storage to **certifications** and trust marks. Early adopter clients might even earn **"Quantum Ready" badges or certifications** to showcase their security posture. Such programs allow a company to monetize its expertise: for instance, offering **quantum-readiness audits, post-quantum cryptography (PQC)** upgrades, or subscription services for quantum-secure communications. These new offerings not only generate revenue directly but also deepen customer loyalty (clients stay for the added security) and attract security-conscious customers who might otherwise look elsewhere.

3. Extended Product Lifespan & Lower Liability: Incorporating quantum-safe measures today can **extend the viable life of products and data** into the quantum era. Technology or devices deployed with classical encryption may face obsolescence or urgent replacement once quantum attacks loom; by contrast, quantum-resistant products remain **trustworthy for longer**, providing long-term value to customers. This future-proofing reduces the vendor's liability and costs over time – think of fewer urgent patches, recalls, or legal settlements due to broken encryption. Avoiding a quantum-induced breach also means avoiding the enormous costs that come with failure. (For perspective, the average cost of a data breach hit **$4.88 million in 2024**, and major breaches like Equifax's have cost an estimated **$700 million** plus a 30% stock value drop.) In contrast, the investment in upgrading crypto and systems to PQC is modest by comparison, and it spares the firm catastrophic loss of customer trust and value.

4. Insurance and Financial Upside: A strong security posture *today* can pay off in very tangible ways. Cyber insurers are beginning to recognize quantum risks – for example, leading underwriters now even offer coverage specifically for **quantum computing exploits**, incentivizing clients to be proactive. Companies that can demonstrate quantum readiness are seen as lower-risk, which can **lower cyber insurance premiums**. Some tech providers report that deploying advanced protections (including post-quantum encryption) has helped clients cut insurance premiums by as much as **40%** due to the reduced breach risk. Moreover, a quantum-secure company may enjoy higher valuations and easier access to capital: investors know that the business is less likely to suffer a massive breach-related loss, and more likely to maintain continuity and growth. In short, **quantum readiness translates to financial resilience** – through direct revenue on the offense, and cost avoidance on the defense.

Readiness as a Signal of Strength

Investing in quantum readiness sends a powerful message to the market. It tells customers, partners, and shareholders that your organization is **resilient, forward-looking, and trustworthy** in the face of emerging risks. This signal can be as valuable as the technical benefits:

1. Winning Trust and Confidence: In an era of data breaches and privacy concerns, demonstrating **quantum-safe practices** bolsters your credibility. Customers and partners take comfort in knowing you are safeguarding data not just for today but for tomorrow's threats. Proactive moves like adopting PQC or quantum key distribution show a *commitment to advanced cybersecurity*, which **improves reputation and trust with partners and clients**. Organizations that "show their work" here – for example, by publicizing quantum-safe encryption of sensitive data or achieving recognized quantum-safe certifications – earn a halo of trust. This can directly translate into sales: many **security-conscious clients** will favor vendors who have **stayed ahead of the competition** on security. Conversely, firms that drag their feet risk appearing negligent; being caught unprepared for quantum threats could trigger reputational loss for having ignored clear warnings.

2. Brand Resilience and Innovation: Embedding quantum readiness into your strategy elevates your brand. It positions the company as not just reacting to change, but **anticipating it**, which is a hallmark of industry leaders. Executives often speak about "resilience" – here is a concrete way to embody it. By upgrading cryptography and processes ahead of time, you are showing that the organization can absorb shocks and continue to protect stakeholders in any scenario. The **World Economic Forum** notes that this kind of strategic shift *"safeguards data and systems while also reinforcing trust and confidence among customers and stakeholders"*. It is evidence of good governance and foresight. Internally, this mindset of innovation can boost morale (teams feel proud to work at the cutting edge), and externally, it strengthens brand equity: your company becomes synonymous with security, reliability, and future-readiness.

3. Investor and Regulator Confidence: Quantum readiness is increasingly viewed as a marker of sound risk management. Cybersecurity has in fact become a **top-tier ESG concern** – nearly two-thirds of institutional investors rank cybersecurity as the #1 ESG issue, on par with climate risk. The rationale is clear: cyber vulnerabilities pose a *"profound, business-wide risk"*, and breaches can

jeopardize customer loyalty and trust, disrupt operations, and even expose companies to legal liability. For boards and investors, a company that is actively mitigating even the next generation of threats demonstrates exceptional governance. This can translate to easier due diligence processes, higher investor appetite, and a premium on your stock or valuation (because the business is less likely to suffer a quantum-related catastrophe). Regulators, too, are starting to prefer (and even mandate) quantum-safe transitions. Governments and standards bodies are urging businesses to prepare for a quantum-safe future – **non-compliance with emerging PQC standards could invite legal and financial repercussions**. Being ahead of these mandates not only avoids penalties but positions your firm as an industry leader shaping the standards, rather than scrambling to meet them.

4. **Outpacing High-Trust Sector Peers:** In sectors where security and trust are paramount – finance, healthcare, pharma, cloud services – quantum readiness can set you apart **when it matters most**. Imagine a pharmaceutical company entrusted with decades of research data or patient records: if it can assure clients that those assets are protected by quantum-resistant encryption, it will have a clear edge in winning partnerships (e.g. hospital systems or biotech collaborators will prefer it over a competitor without such guarantees). In financial services, we already see big players treating quantum as a strategic priority. Banks like JPMorgan, HSBC, and Goldman Sachs have dedicated quantum research teams exploring both **quantum computing applications and quantum-safe security**, knowing that this dual focus will strengthen customer trust and their innovative image. A global bank that understands **quantum's implications for cybersecurity and customer trust** is better positioned to attract clients (and deposits) than one that ignores the coming shift. Similarly, a cloud or SaaS provider that builds quantum-safe encryption into its platform can become the default choice for government and enterprise contracts that demand long-term data security. In short, **leaders in quantum readiness are poised to grab market share** from slower rivals, especially in high-scrutiny industries where every trust advantage counts.

New Opportunities and Markets Emergence

Quantum readiness isn't only about managing risk – it's also about *seizing new opportunities*. As quantum technologies mature, entirely new markets and revenue streams will develop for those prepared to capitalize:

1. Quantum-Driven Insights and Solutions: The same quantum revolution threatening cryptography is also unlocking unprecedented computing power. Companies that invest early in **quantum computing exploration** can gain capabilities that translate into superior products and services. Consider high-value problems like complex optimization, simulation, and machine learning – quantum computers promise breakthroughs here. For example, financial institutions are experimenting with quantum algorithms for **risk modeling, portfolio optimization, and fraud detection**, tackling computational challenges in new ways. Pharmaceutical firms are partnering with quantum tech companies to accelerate drug discovery – **quantum simulations of molecular interactions** could *"dramatically reduce the time and cost to bring new therapies to patients"*. These are not sci-fi scenarios but active pilot projects in 2025. By being quantum-ready (in talent, partnerships, and infrastructure), your business can be among the first to *offer* quantum-enhanced solutions in your industry. This might mean new products (e.g. ultra-optimized supply chain routes, AI models boosted by quantum computation, materials designed with quantum chemistry) that competitors without quantum capability simply can't match.

2. Partnering in Quantum Ecosystems: No company can develop the quantum future alone – the winners are joining forces in the emerging **quantum ecosystem**. Being quantum-ready increases your attractiveness as a partner to both tech providers and fellow industry leaders. A case in point: more than **200 organizations, including Fortune 500 firms, start-ups, and research labs, have joined the IBM Quantum Network** to share knowledge and access quantum computing resources. By engaging in such networks or consortia, your company gains early insight into breakthroughs and can co-develop practical applications. We see telecom operators, banks, automakers, universities – entire value chains – coming together to pilot quantum use cases. **Partnerships in the quantum era** can spark new business models (for instance, offering quantum-powered services via the cloud, or integrating quantum-secure communications for customers). If you're quantum-ready, you can secure a seat at the table for these multi-company innovations. In turn, that positioning can lead to first-mover advantage in new markets that emerge when quantum computing achieves practical utility.

3. Advisory and Enablement Services: An often overlooked win:

turning your internal quantum expertise outward. If your organization masters quantum-safe security or develops quantum computing know-how, you can **offer this expertise as a service**. This could be direct – such as a consulting arm that advises other companies on how to upgrade their encryption (monetizing your hard-earned readiness playbook) – or indirect, like packaging your quantum-safe technologies for sale. For example, a software firm that rebuilds its platform with PQC might spin off a toolkit for developers to do the same. A bank that built quantum-risk assessment tools could license them to smaller banks. Even providing **training and certification** in quantum security to your supply chain or customers can deepen relationships and potentially be a chargeable offering. We're already seeing large tech companies do this: Microsoft's new **Quantum Ready program** is designed to **provide business leaders with tools and workshops to become quantum-ready** – effectively creating an ecosystem of partners and clients who will rely on Microsoft's quantum technologies in the future. Your company, too, can become a recognized leader or **enabler in the quantum space**, opening up advisory revenue and reinforcing your brand's authority.

4. New Market Entrant Advantage: In some cases, quantum readiness can let you *enter entirely new markets.* For instance, a logistics provider that develops quantum-optimized routing algorithms might move into offering a premium global supply chain service that outperforms competitors on cost and speed (thanks to quantum calculations). A healthcare company that secures patient data with quantum-safe methods might attract international clients (e.g. European partners post-GDPR) who demand the highest security, effectively entering markets that were closed to less secure competitors. By aligning your R&D with quantum advancements, you position the firm not just to protect what you have, but to **expand into new domains** as opportunities arise.

The Business Case: Costs, Returns, and Strategic Positioning

Every executive will rightly ask: *what's the ROI on quantum readiness?* The answer becomes clearer when you frame it correctly – it's not simply an "insurance policy" cost, but a strategic investment that yields both **offensive and defensive returns**. Let's break down the business-case considerations:

- **Cost of Readiness vs. Cost of Inaction:** On one side of the ledger is the investment to become quantum-ready –

upgrading systems to PQC, auditing cryptography, training staff, perhaps engaging experts or replacing hardware over time. These costs are real, but manageable and spread over years. Now consider the **cost of a quantum-induced breach** if you do nothing: the moment a capable quantum computer emerges, all data you secured with old encryption becomes fair game. The financial impact could be devastating – imagine confidential customer records, intellectual property, or transaction logs being exposed. We already know classical breaches are extremely costly (average **$4M–$5M** each, with mega-breaches like Equifax reaching **hundreds of millions**). Multiply that by the scale of encrypted data at risk across years, and add the **intangible costs** of lost trust, regulatory fines, lawsuits, and business disruption. Suddenly, the upfront spend on quantum-safe crypto looks like a rounding error. In business terms, quantum readiness has a high **risk-adjusted ROI**: it averts a potentially existential risk (the "loss avoided" is enormous) while also delivering the positive returns discussed (new revenue and brand lift).

- **Using Readiness in Sales and Contracts:** Quantum readiness is quickly becoming a **selling point** and could soon be a requirement in RFPs and due diligence. Forward-looking organizations are already including their security roadmap in proposals – for example, a cloud vendor responding to a major bank's RFP might highlight that they use NIST's PQC algorithms for all data encryption, thereby guaranteeing that the bank's data will remain safe for the contract's duration (and beyond). This can **tip the scales in competitive bids**. We can anticipate procurement questionnaires starting to ask suppliers: *"Do you have a plan for post-quantum security?"* Being able to check "Yes – already implemented" might be the difference between winning or losing a multi-million dollar deal in the future. Similarly, in mergers and acquisitions, acquirers are beginning to scrutinize the target's cyber posture. A company deeply dependent on legacy encryption might carry unquantified **quantum risk debt**, whereas a quantum-prepared company is a safer bet. Thus, your readiness can **preserve or enhance company valuation during M&A** (no sudden discounts for security gaps) and smooth out partner due diligence processes.

- **Regulatory and Compliance Alignment:** As mentioned, regulators are pushing for action. Standards are emerging (e.g., NIST's new PQC standards like FIPS 203/204/205 finalized in 2024) and governments have set timelines for their agencies to adopt quantum-safe encryption. This trend will trickle down to the private sector through critical infrastructure regulations, data protection laws, and industry standards. By moving early, you **avoid the scramble and potential penalties** of future compliance deadlines. Moreover, you get to help shape the narrative – possibly even influencing standards in your favor or gaining a seat on industry bodies setting the rules. In terms of **ESG and corporate governance**, being proactive here is a strong positive signal. It shows your enterprise not only complies with current regulations but is *ahead of them*, embodying the principle of doing what's right before being told to. This alignment with the "Trust" and "Governance" aspect of ESG can attract investors and clients who prefer stable, well-governed partners.
- **Intangible Benefits – Trust, Brand, Talent:** Some returns are harder to put in a spreadsheet but are nonetheless significant. Quantum readiness can be **marketed as part of your brand story** – in press releases, annual reports, customer marketing, you can highlight your commitment to cutting-edge security ("We are among the first in our industry to implement quantum-resistant encryption, ensuring your data remains safe with us for decades to come"). This builds confidence and can be a tiebreaker for a reputation-conscious client. Trust, once lost, is painfully expensive to regain (if it can be at all) . By investing in trust now, you avoid paying the price of trust repair later. Additionally, being involved in exciting areas like quantum technology can help in **recruiting and retaining talent**. Top engineers and analysts want to work at innovative firms; by publicizing your quantum initiatives, you signal that yours is a company of the future, where ambitious talent can grow. This can reduce costs associated with hiring or turnover and spur an internal culture of innovation that yields new ideas beyond quantum.

In summary, the business case for quantum readiness stands on multiple pillars: preventing massive downside, creating upside opportunities, and strengthening the very foundations of trust and value that your business runs on. The cost is justified not by fear, but by strategy – it's an investment in staying **ahead of the curve** and reaping the rewards of being there first.

Executive Signal: *Are you marketing your quantum advantage to customers and partners yet?*

This question is a final call-to-action for leadership. If you've taken steps to be quantum-ready, **make it known.** Use it in your marketing and conversations with stakeholders. Ensure that your customers understand the added value you provide by future-proofing their data. Let partners and investors know that your foresight is part of what makes you a reliable, long-term winner. And if you haven't started on this journey, the time to begin is now. Quantum readiness is not just a technical necessity; it is fast becoming a **business advantage** that distinguishes the leaders from the laggards. In the race to *Be Ahead, Be Safe, Win*, those who turn readiness into strategy will ultimately come out on top.

PART IV
THE EXECUTIVE PLAYBOOK

This final section translates strategy into action. It gives you a practical 12-month plan, a vision of the road ahead, and the manifesto that will guide your organization to thrive in the quantum era.

8 YOUR NEXT 12 MONTHS THE EXECUTIVE QUANTUM READINESS PLAN

Why the Next 12 Months Are Critical

The coming year represents a **critical window** for getting ahead of quantum risk. Government standards for post-quantum cryptography (PQC) are emerging now, and adversaries aren't waiting – they are actively **harvesting encrypted data today to decrypt later** with quantum tools. Leaders cannot assume they have a decade of slack; in fact, **2035 is viewed as a "finish line," not the starting line** for full PQC migration. Large organizations may need 5–10 years to fully transition, so delaying preparation even a year could put you behind schedule. Regulators have taken note: for example, the U.S. Office of Management & Budget (OMB) gave federal agencies only 6 months to inventory their cryptographic systems, and **NSA requires certain critical applications to use quantum-resistant solutions by 2025**. In the EU and Canada, similar PQC mandates are accelerating timelines. In short, the **time to act is now** – waiting for "quantum day" to arrive will be too late.

Beyond risk, there's a **competitive advantage** in being proactive. Early adopters of quantum-safe practices send a powerful signal to customers and partners that their data will remain safe for the long term. This builds trust and can satisfy emerging compliance expectations ahead of peers. Conversely, a slow response could mean playing catch-up under regulatory scrutiny or losing customer confidence. **Data has longevity** – sensitive records from healthcare to financial contracts often remain confidential for years or decades. Any data still secret five or ten years from now is *already* at risk if encrypted with legacy algorithms. By acting in the next 12 months, executives can **get in front of this risk**, ensure their organization's encryption won't become tomorrow's weak link, and seize a leadership position. As one cryptography expert put it, *"the only safe option is to start immediately"*.

Quick Wins in the First 90 Days (0–3 Months)

In the first quarter of the plan, focus on **laying the groundwork** and scoring a few quick wins to build momentum and buy-in:

- **Appoint an Executive Sponsor and Team:** Designate a C-suite **quantum lead** (or "champion") and assemble a cross-functional **crypto transition team**. This team should include IT/security, risk/compliance, and business unit representatives. Give it executive visibility and authority – PQC readiness *cannot* be an afterthought buried in IT. As NIST's Dustin Moody emphasizes, establishing a dedicated team with leadership backing is the essential first step. Make roles clear (e.g. a project lead, cryptography specialist, vendor liaison) and set a meeting cadence.
- **Inventory Your Cryptography:** Kick off a **comprehensive cryptographic inventory** across the organization. Most organizations do not fully understand where all their encryption and digital signatures are implemented, so this is priority #1. Catalog every system, application, database, product, and third-party service that uses cryptographic algorithms (especially RSA, ECC, Diffie-Hellman, or any other quantum-vulnerable schemes). This "crypto asset inventory" is the foundation for everything to come. Leverage automated discovery tools to scan code and network traffic for known algorithms, and send questionnaires to vendors for any black-box systems. The inventory should note what algorithms are used and **which data or functions they protect**, as well as an estimate of how long that data needs to remain secure. These details will inform risk prioritization. *Quick win:* Even if the inventory is not 100% complete in 90 days, having an initial **Cryptographic Bill of Materials (CBOM)** for major systems is a tangible deliverable to show progress.
- **Identify a Pilot Use-Case:** Select one target system or

application as a **pilot for quantum-safe technology**. It should be important enough to matter (e.g. a customer-facing web service or an internal PKI system), but not so mission-critical that experimenting would be high risk. The goal is to have a contained environment to **test PQC algorithms or integrations** in the coming months. For example, you might pick an internal tool and plan to implement a post-quantum encryption library in its next update, or enable a *hybrid* (classical + quantum-safe) TLS option on a non-critical service. Choosing a specific pilot early forces the team to start learning by doing. Industry experts recommend *"starting with one specific area to explore and experiment with migration"* as a learning exercise. By the end of 90 days, you should have the pilot project defined, resourced, and scheduled.

- **Raise Awareness and Train the Core Team:** Use this kickoff period to begin **educating stakeholders** and training the implementation team on quantum basics. Hold an executive workshop to brief leadership on why quantum readiness matters, aligning on the 12-month roadmap. Simultaneously, ensure your IT and security staff get up to speed on new PQC algorithms (e.g. through NIST's latest standards documentation) and crypto-agility concepts. Consider a short internal awareness campaign demystifying quantum threats for broader staff. This investment in **training and awareness** pays off by creating a shared urgency without hype. It can be as simple as lunch-and-learn sessions or sharing digestible briefs about the quantum threat landscape and upcoming changes in encryption standards. Early team education also fosters a culture of continuous learning, which you'll need as standards evolve.
- **Low-Hanging Fruit ("Vendor Pressure"):** In the first 90 days, an easy win is to **engage your key technology vendors** in conversation about their quantum-safe roadmap. Reach out to your cloud providers, software vendors, and hardware suppliers with a standard set of questions: "Are you working on PQC support? When will it be available? Do you have a transition plan for customers?" This not only signals to vendors that your organization expects action, but also lets you **document vendor commitments (or lack thereof)**.

CISA recommends proactively asking vendors for their quantum-ready roadmaps, including timelines for testing and integrating PQC into their products. If a critical vendor has no good answer, that's a red flag you can escalate (and it may influence future procurement or contract renewal decisions). Simply *asking* these questions early is a quick win – you'll compile a list of which partners are ahead or behind on PQC. This informally applies pressure on laggards and gives you input for risk planning.

By the end of the first 3 months, you should be able to report a few concrete achievements to the board: an executive-sponsored program is in place, an initial cryptographic inventory is underway (with initial findings documented), a PQC pilot project identified, and vendor discussions started. These **quick wins** set the tone that quantum readiness is actionable and being treated with urgency.

Executive 12-Month Quantum Action Roadmap

Months 1–3	Months 4–8	Months 9–12
Immediate Actions	Mid-Term Initiatives	Future-Proofing
Quick Wins	Operational Changes	Long-Term Readiness
Audit current encryption usage	Pilot PQC solutions	Implement PQC in critical systems
Executive & team education	Establish crypto-agility governance	Review emerging quantum tech opportunities
Push vendors on quantum-safe options		

6–12 Month Milestones (Building Momentum)

After the initial groundwork, the next 6–12 months are about **executing on the roadmap** and achieving several key milestones:

- **Complete the Cryptographic Risk Assessment:** With your inventory in hand, conduct a **quantum risk assessment** to map out where you are most vulnerable. Not all crypto usage is equally critical. Identify which systems and data would cause the greatest damage if compromised via quantum attack – for instance, customer PII, sensitive intellectual property, financial transactions, authentication systems, etc. Assess the "shelf life" of the data each system protects (data that must stay confidential for 5+ years is highest priority for quantum-safe upgrades). The output should be a **Crypto Risk Map** that ranks systems/data by priority. This will guide what gets migrated first and which areas might justify more aggressive timelines or resources. If the inventory is the *what* (what do we have), the risk map is the *so what* (what do we tackle first and why). In practice, this means overlaying your crypto inventory with business impact: e.g., highlight any High-impact or regulated systems using RSA/ECC, and evaluate the feasibility of upgrading them sooner. Regulators expect agencies and enterprises to **prioritize high-value assets** in their quantum transition plans. Aim to finish an initial risk assessment by around the 6-month mark. Even if it's not perfect, it will bring clarity to where to focus remediation and pilot efforts. (Keep updating this as you discover more systems.)

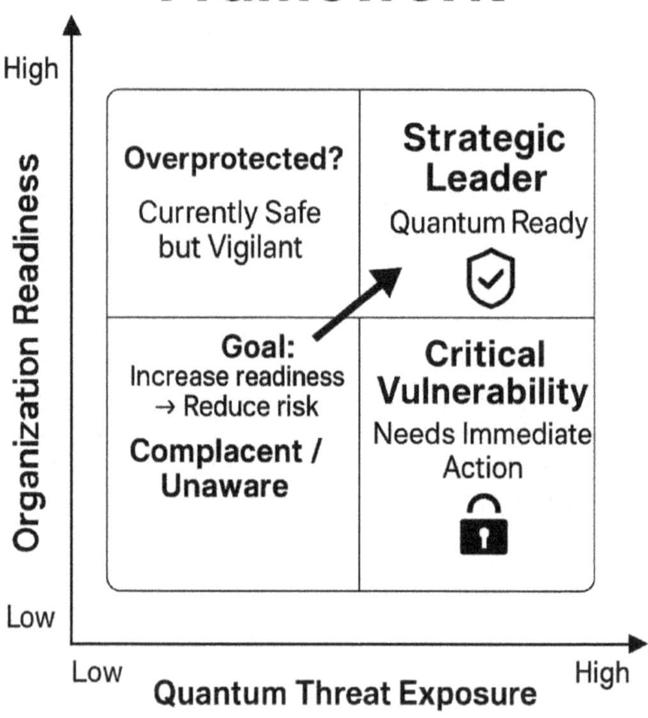

- **Pilot a Post-Quantum Solution:** By month ~6, your chosen pilot project should be underway – if possible, even **tested or implemented with a PQC algorithm** in a controlled environment. This could mean running a prototype using one of NIST's draft standard algorithms (like Crystals-Kyber for key exchange or Crystals-Dilithium for signatures) in the pilot system and observing performance and compatibility. The outcome of the pilot will teach invaluable lessons about integration challenges. Did the new algorithms impact latency or throughput? How did they interact with existing protocols? Document these results. By 9–12 months, the pilot **should be completed** and its findings feeding into your broader plan.

Early pilots are happening across industries – for example, some organizations have started *piloting quantum-resistant algorithms in non-critical systems* to gain experience. The goal is that by the 1-year point, you can say: *"We have tested PQC in one of our systems and proven it works (or identified what needs fixing)."* This tangible progress moves you from theory to practice, and positions you to expand pilots afterward.

- **Implement a "Crypto-Agility" Demonstration:** In parallel with the pilot, strive to **demonstrate crypto-agility** on a small scale by the end of the year. *Crypto-agility* is the ability to swap out cryptographic algorithms with minimal disruption. It's considered essential for the quantum era, since algorithms and standards will continue to evolve. For your organization, a crypto-agility proof-of-concept might involve updating a component (say, an API service or an IoT device) so that it can use either the legacy algorithm or a new PQC algorithm interchangeably. This could be as simple as modifying a module in a software application to pull the cryptographic algorithm type from a configuration, allowing easy updates. The point is to **prove that your systems can adapt**. If your pilot already includes some agility features (e.g. a dual-stack implementation), great – if not, consider a separate mini-project to refactor one component with a modular crypto interface. This exercise will highlight where hard-coded assumptions or rigid protocols exist. By month 12, aim to have at least one concrete example of crypto-agility in your tech stack (even if just in a test environment) that you can show leadership: *"Here's a dashboard where we can flip from RSA to Kyber in real-time without service downtime"*. That demonstrates future resilience. Crypto-agility isn't just a buzzword; NIST and others emphasize it as critical design goal for new systems. Getting a jump on it now will pay off when more PQC options emerge.

- **Evaluate Hybrid Cryptography Options:** As you progress, evaluate whether **hybrid cryptography** makes sense for your interim strategy. Hybrid means using a classical algorithm (like RSA/ECC) *and* a PQC algorithm together (e.g. in a TLS handshake) so that an attacker would need to break both. This approach can provide extra assurance during a transition period. Some vendors and open-source libraries already offer hybrid modes. However, be mindful of complexity – combining algorithms can introduce new vulnerabilities if not carefully implemented. Use this year to research and perhaps test a hybrid approach in non-production settings. By month 9 or so, you should have a position on hybrid solutions: Will you adopt them as a bridge (for example, to satisfy a compliance requirement in the next 1-2 years), or will you leapfrog directly to full PQC for most systems? Dustin Moody at NIST notes that hybrid deployments are "conservative" and can buy time, but also carry their own risks. The decision will depend on your risk appetite and timeline. Either way, having **evaluated hybrid vs. pure PQC** by year's end makes your roadmap more concrete. If you decide to use hybrid encryption in some places (say, upgrading VPN or TLS configurations), that could even become **another pilot** to execute in the next phase.

- **Engage and Align Vendors and Partners:** By 6–12 months in, you should have deeper engagement with vendors beyond the initial queries. **Gather responses and roadmaps** from each critical supplier and create a consolidated view: e.g., "Out of our 10 key software platforms, 6 have confirmed they will support PQC by 2024, 3 are evaluating, 1 has no plan yet." This vendor PQC status report will inform your risk planning (systems whose vendors are lagging might need replacement or compensating controls). It's also a good time to push for **contractual commitments**: as you renew contracts or service agreements, insert requirements for quantum-safe encryption updates. CISA advises organizations to proactively plan for contract updates ensuring older products get PQC upgrades or new purchases have PQC "built-in" by default. If you operate in a supply chain, coordinate with partners on quantum-safe data exchange plans (for instance, if you send encrypted data to a supplier, ensure *they* are planning for PQC too). By the end of the year, vendor engagement should move from asking questions to **establishing joint timelines**. The ideal milestone: you've secured at least one or two **vendor commitments in writing** (e.g. a product's upgrade schedule or a formal statement of PQC support). This reduces uncertainty going forward.

- **Refine the Roadmap and Governance:** As these initiatives progress, solidify your ongoing **governance structure** for quantum readiness. That may mean integrating the quantum readiness team into existing risk governance, setting up a quarterly board report (more on that below), and planning beyond the 12-month horizon. By month 12, the team should deliver an updated **Quantum-Readiness Roadmap** document to leadership, incorporating everything learned. This roadmap is effectively your multi-year plan (e.g. "2025: pilot and inventory phase; 2026–2027: migrate top 5 systems; 2028: transition 50% of systems…" and so on, aligned with any regulatory deadlines). Include an **updated risk register** for quantum threats and any policy changes needed (like new encryption standards for all development projects starting next year). Also consider if a formal **"crypto center of excellence"** or permanent crypto officer role is warranted to own this issue going forward. By treating the next 12 months as a pilot year for your *process*, you'll be much better positioned to scale up efforts in year 2.

Budgeting for Quantum Readiness

What will all this cost, and what should you budget for in year one? Given that the U.S. federal government estimates spending **$7.1 billion on PQC migration over the next decade**, private-sector organizations should expect non-trivial investments as well. The good news is that many quantum-readiness steps (inventory, asset management, crypto upgrades) overlap with *good cybersecurity hygiene*, and can often be funded as extensions of existing security initiatives. Here are the key budget considerations for the next 12 months:

- **Talent and Training:** If you lack internal cryptography expertise, consider hiring or contracting a **cryptographic specialist** to support the team. At minimum, allocate part of an FTE (or consultant hours) for someone who deeply understands encryption and can evaluate PQC libraries, protocols, etc. Additionally, budget for **training programs** so your existing staff can get up to speed on post-quantum algorithms and tools. This might include sending team members to a NIST PQC workshop, taking an online course on quantum computing fundamentals, or attending industry conferences. Investing in talent not only helps execution but also aids retention – top cyber professionals are drawn to organizations tackling cutting-edge challenges. In short, ensure funds for both **specialized expertise** and **ongoing education**. This could be on the order of a few hundred hours of expert consulting and several training courses/conferences in the first year.

- **Tools and Technology:** Plan for some **tooling expenditure** to accelerate your cryptographic inventory and migration. For example, you might purchase a **crypto discovery tool** that scans code repositories for vulnerable algorithms, or an upgrade to your certificate management system to handle new algorithm types. You may also need to update hardware security modules (HSMs) or other cryptographic hardware to support PQC – some vendors already offer HSM firmware supporting certain new algorithms. Include budget for **testbed environments** as well: spinning up cloud instances or lab systems to pilot PQC will incur minor costs. Make sure to provide your team with "**state-of-the-art tools** for cryptographic analysis, data discovery, and systems inventory," as one best-practice guide advises. Don't forget software licensing costs if any of the new libraries or vendor product updates require it (some PQC libraries are open source, but enterprise-grade solutions or support might not be). The first year is likely more about assessment and pilot tools than full-scale deployments, so budget accordingly (possibly six figures for tools across a larger enterprise, less for a smaller firm).

- **Testing and Pilot Budget:** Your pilot implementations and any crypto-agility proof-of-concepts will require **time and resources to test thoroughly**. Allocate budget for developers/engineers to spend extra cycles on refactoring and testing systems with new algorithms. Performance testing is especially important – you may need to run parallel environments to measure impact of PQC vs. legacy encryption (for instance, testing transaction throughput or latency differences). This might involve contracting a third-party testing firm or using cloud test services to simulate load. Ensure your plan includes funding for these **evaluation projects** – essentially R&D expenditure to shake out issues before wider deployment. It's much cheaper to find out in a pilot that Algorithm X is too slow for your needs than to discover it after deploying broadly. Thus, **budget for**

experimentation. Some leading organizations treat this as an innovation budget line item.

- **Vendor Support and Transitions:** As you engage vendors, you may find that getting quantum-safe upgrades requires **updating to newer product versions or licensing additional features**. Be prepared for some unplanned spend here. For example, your VPN appliance vendor might only offer PQC algorithms in the latest hardware model – which means you might have to budget for an earlier hardware refresh than expected. Or a software vendor might put PQC in an "enterprise plus" tier. While you should certainly negotiate and push vendors to include security upgrades at minimal cost, the reality is there could be **capital expenses for new equipment or software** in the coming 1–2 years directly tied to quantum readiness. Identifying these in the first 12 months means you can plan in advance. Additionally, consider budgeting for **vendor management efforts** – you might need to fund a small **contingency for replacing a non-compliant vendor** if one refuses to address PQC (e.g. allocate budget to pilot an alternate solution). Finally, if you use external consultants or auditors to validate vendor claims (for supply chain assurance), include those service fees.

- **Contingency for Upside Opportunities:** On the flip side, keep a slice of budget aside for **unexpected opportunities** – for instance, if a promising new PQC solution or library emerges mid-year, or if a regulatory grant/funding becomes available to subsidize certain upgrades. Quantum tech is a fast-moving field; being able to allocate some funds quickly to a new tool or collaboration (say, joining an industry consortium pilot) could provide outsized benefit. An agile budget will help you stay ahead.

One strategic angle to note: **Quantum readiness can help unlock budget from the boardroom.** Framing these investments as preparing for inevitable compliance mandates and protecting long-term value makes them easier to justify. In fact, boards and regulators increasingly want to see "concrete progress within months, not years" on PQC. Tie budget requests to both **regulatory drivers** (e.g. upcoming laws, industry standards) and **co-benefits** (like improved crypto inventory and cyber hygiene overall). Often, you can fold quantum-related funding into existing IT refresh budgets or security upgrades by highlighting the overlap. The bottom line is to **secure enough budget for tools, training, and testing in Year 1** so that the program doesn't stall. A well-resourced program – with dollars attached to it – signals to everyone that leadership is serious about meeting quantum threats head-on.

Reporting Progress to the Board

Quantum readiness should be elevated as a **board-level agenda item** within the year. Executive leadership will need to update the board (or relevant risk committee) on progress, challenges, and next steps. Here's how to communicate and measure your quantum program in a way that keeps the board engaged:

- **Define Metrics and a Readiness Dashboard:** Establish clear **metrics for post-quantum readiness** and track them over the 12 months. For example, metrics might include: *% of cryptographic assets inventoried*, *% of high-risk systems mitigated or scheduled for mitigation*, *number of PQC pilots completed*, *vendor PQC readiness status (e.g. 3 of 5 critical vendors deliver PQC support by Q4)*, and *training metrics* (staff trained, etc.). Consider creating a simple **dashboard or scorecard** that visualizes these – for instance, a traffic-light status for inventory completeness, a gauge for "quantum readiness score" (which could be an aggregate of various factors), and a timeline of key milestones achieved. The goal is to **quantify progress** and highlight momentum. Boards respond well to numbers and trend lines. If at month 3 you report "30% of known cryptosystems inventoried" and by month 6 it's "80% inventoried, 2 systems migrated in pilot," that demonstrates concrete action. Over time, you might develop a **Quantum Readiness Index** for your organization – even if informal – to rate your posture (e.g. 7/10 now, aiming for 9/10 next year). Importantly, tie these metrics to business risk: for example, "After our inventory, we determined 5 high-risk applications. Today, 2 are in pilot for PQC – meaning roughly 40% of our high-risk data will be quantum-safe by year-end." This kind of messaging resonates at the board level.

- **Integrate with Enterprise Risk Reporting:** Emphasize that quantum risk is being managed as part of the company's broader **enterprise risk management (ERM)** and compliance obligations. Boards are increasingly asking about emerging risks, and quantum falls squarely in that category. In reporting, position your quantum readiness program as *not just an IT project*, but a core risk mitigation initiative akin to other strategic risks. For example, if your board materials include a cybersecurity risk section, include a subsection on **"Quantum Cryptography Risk"** with a rating (e.g. moderate today, trending lower with our actions). Highlight any **material risks or disclosures**: if you foresee needing to disclose quantum readiness in financial filings or to regulators, mention that and how you're addressing it. Also, note that regulators/insurers are starting to ask questions (some regulators already ask "Do you know where your quantum-vulnerable systems are?"). Showing the board that you're ahead of those questions builds confidence.

- **Demonstrate Long-Term Preparation:** Boards care about avoiding surprises and legal liabilities. Communicate how your 12-month progress sets the foundation for long-term **trust and compliance**. For instance, update any **data retention or encryption policies** and inform the board that policies have been adjusted in light of quantum risk (e.g. "We've updated our data retention and encryption policy to mandate PQC for any data needing confidentiality beyond 2030"). If your industry has specific guidelines or if customer contracts start to include PQC requirements, brief the board on those as well. The key message: *we are proactively addressing this and will not be caught flat-footed.* You can reinforce this by referencing external benchmarks: for example, mention that *"Only ~5% of organizations have started planning for quantum risk"* (per an ISACA study), and thus your company is ahead of most peers – turning preparedness into a competitive differentiator.
- **Showcase Customer/Regulatory Posture Improvements:** If part of your business involves customer assurances (e.g. in B2B contracts or public trust), highlight how your quantum-safe journey is strengthening that posture. For instance, if major clients have inquired about your encryption or if you can market your organization as "quantum-ready," mention this strategic benefit to the board. Perhaps you can say, "Our proactive quantum upgrade plan has been noted positively by key customers/regulators" – even better if you have an example, like a customer security

questionnaire that you could answer confidently due to your inventory and roadmap. As quantum-safe standards become a selling point, the board should view this program as enhancing the **brand and customer trust**. On the regulatory side, if you're in a regulated sector, map your metrics to any compliance mandate (for example, if a law requires a cryptographic inventory or transition plan by 2024, report that 'we have completed X% of that requirement'). The board dashboard can include a "regulatory compliance" indicator for quantum readiness specifically.

- **Regular Updates and Escalation:** Plan to **report to the board or a committee at least quarterly** on quantum readiness throughout the year. Start with an initial briefing (what the risk is, why the 12-month plan, what the quick wins are) – many board members will appreciate even the educational aspect. Then provide a progress update mid-year and at year-end. Keep these updates high-level and solution-focused: e.g., *"Here's our quantum risk posture score as of Q2, here were our Q2 milestones achieved (inventory done, pilot launched), and here's what's next."* Use visuals from your dashboard to make it skimmable. Additionally, don't shy away from **escalating roadblocks**: if, say, a critical vendor has given a poor response and represents a risk, flag that to the board (along with mitigation steps, like exploring alternatives). A well-informed board can help remove obstacles – for instance, they might approve additional funding or policy changes to address the issue. By treating the board as a partner in this journey, you also ensure ongoing executive support. In summary, convert quantum readiness from a niche technical concern into a regular boardroom discussion about risk management and strategic preparedness.

12-Month Timeline: Month-by-Month Priorities

A sample 12-month timeline for executive quantum readiness. It highlights quick wins in the first quarter (cryptography inventory completed, sponsor/ team appointed, pilot identified), mid-year progress (PQC pilot tested, quantum risk assessment completed, vendor roadmaps reviewed), and end-of-year outcomes (first system made quantum-safe and a board readiness report delivered). This visual timeline aligns key actions to each milestone, ensuring stakeholders have a clear roadmap of what to achieve by Month 3, 6, 9, and 12.

To further clarify the roadmap, here is a **month-by-month view** of action priorities, broken into approximate quarters:

- **Months 1–3:** *Foundation and Quick Wins.* Stand up the program (executive sponsor named, team formed), complete the initial cryptographic inventory, and deliver an **inventory report** identifying all major cryptographic systems. Choose the pilot project and procure any tools or training needed. Engage vendors with initial queries. By the end of Q1, you should have momentum – the board and organization know a quantum readiness initiative is underway and tangible first steps are done.
- **Months 4–6:** *First Implementation and Risk Analysis.* Begin the PQC pilot implementation in the selected system (by Month 6 you should be testing the new algorithm in that environment). Simultaneously, analyze the inventory to produce the **crypto risk heatmap** – pinpoint the highest-risk systems/data. Conduct a formal risk assessment workshop to prioritize systems for migration. Start drafting internal policies or guidelines for new cryptography (e.g. a policy that all new projects must consider PQC by design). By the end of Q2, ideally the pilot test is *completed* or nearly so, the risk assessment is complete, and you have engaged with vendors enough to know their roadmaps. Present a **mid-year report** to the board with these findings and any adjustments to strategy.
- **Months 7–9:** *Broaden and Refine.* With pilot results in hand, plan the next phase: perhaps schedule the first production roll-out of a quantum-safe algorithm (if the pilot was successful) for a high-priority system. Work on the **crypto-agility proof-of-concept** – this could be done in a lab environment by Month 9. Evaluate hybrid encryption options through research or a small-scale test. By this time, you should also be updating your **incident response and business continuity plans** to account for quantum threats (for example, adding scenarios about "what if encrypted data is suddenly compromised?"). Continue engaging the board; at this stage, you might brief them on budget needs for next year (if scaling up efforts) and any policy decisions (like dropping a non-compliant vendor or accelerating a certain upgrade). Aim to have **vendor commitments** or at least a clear picture of vendor timelines by end of Q3, so you can incorporate that

into planning.

- **Months 10–12:** *Deliver and Strategize Next Steps.* In the final stretch of the year, focus on delivering an **end-of-year quantum readiness report** and solidifying long-term strategy. Ideally, by Month 12 you will have at least **one production system running quantum-safe encryption** (even if just a subset or in parallel mode) – this checks the box on the oft-asked question "When will our first system be quantum-safe?". Additionally, finalize documentation: updated encryption standards for the organization, a roadmap for the next 2–3 years, and integration of quantum readiness into enterprise risk management. Conduct a lessons-learned session with the team. Celebrate the wins – you've likely improved overall security posture along the way. Present to the board a **"Year 1 Quantum Readiness Scorecard"**: what was achieved (e.g. inventory 100% complete, 1 pilot done, X% of critical systems planned for upgrade), current quantum risk posture (residual risk), and what the plan is for Year 2. If required, also prepare communications for customers or regulators about your quantum-safe progress (this could be a paragraph in your annual report or a note in customer security updates). By the end of the 12th month, the goal is that your organization is **quantum-ready at the strategic level** – you know your risks, you've tested solutions, you have a plan, and you can confidently say you are ahead of the curve in addressing the quantum threat.

Throughout this timeline, maintain agility. If standards or threat intelligence change (e.g. a new PQC algorithm is standardized or a breakthrough happens sooner), be ready to adapt. The 12-month plan should be a living project, but **having this month-by-month discipline prevents inertia**. It creates urgency and accountability which are exactly what's needed for an executive-driven program.

Executive Signals: Key Questions to Guide Your Team

As an executive, asking the right questions sets the tone and ensures focus. In steering your organization's quantum readiness over the next year, consider regularly posing these **"executive signals"** to your team and vendors:

- **"Do we know where *all* of our cryptography lives?"** –

This question should be asked early and often. It reinforces the importance of the cryptographic inventory. You want to hear that the team has a handle on every major instance of encryption or digital signing in the enterprise (and a plan to find the stragglers). If the answer is ever uncertain, it's a cue to dig deeper or allocate more resources, because blind spots in cryptography = unknown risk.

- **"When will we have our *first* system go quantum-safe?"** – This drives urgency for tangible progress. It prompts the team to target a specific system and timeline for implementing PQC. The ideal answer might be, "By Q4 of this year, System X will be running a quantum-safe algorithm in production." By continually asking this, you ensure the program doesn't get stuck in analysis; it must deliver a real outcome that stakeholders (and possibly customers or regulators) can see.
- **"What's our quantum risk *score* today, and how is it trending?"** – Push the team to quantify the risk posture. Whether it's a self-devised score or number of vulnerable systems, this helps track improvement over time. If the trend isn't improving quarter by quarter, you'll want to know why.
- **"Are our vendors and partners keeping up with us?"** – Use this to keep pressure externally. If a critical supplier is lagging, the team should be escalating that to you with options (e.g. pressuring the vendor's executives, or considering alternate solutions). This question signals that you expect the entire ecosystem to move toward safety, not just your own enterprise.
- **"How are we validating our readiness (e.g. drills or audits)?"** – Encourage the use of **tabletop exercises** or third-party audits to test the assumptions. For example, a tabletop exercise could simulate a scenario where an encrypted database is stolen – does the incident response plan account for quantum decryption risk? Asking about this ensures the plan isn't just on paper but is being pressure-tested.
- **"How will we know we're fully quantum-safe, and what could change that timeline?"** – This forward-looking query pushes the team to define the end-state (perhaps "All critical data is protected by PQC or quantum-proof measures by

20XX") and also consider dependencies (like standardization updates, or new tech discoveries). It prepares everyone for the reality that quantum readiness is an evolving target, not a one-and-done project.

In summary, by asking these pointed questions, you drive accountability and clarity. They serve as leadership checkpoints to make sure the *Executive Quantum Readiness Plan* stays on course over the 12 months. Remember, **"This is not a textbook – it's a playbook"**. The next year of effort will position your organization to **Be Ahead, Be Safe, Win** in the quantum era. Your proactive leadership now is what will ensure that, when the quantum computing wave hits, your company is ready to ride it rather than be washed away. The clock is ticking – the executive playbook is in your hands. Let's execute it.

9 LOOKING AHEAD STRATEGIC OPPORTUNITY BEYOND SECURITY

Quantum technologies – spanning quantum computing, sensing, and communication – promise to revolutionize industries and create entirely new opportunities for growth. While much attention has focused on the security implications (e.g. breaking encryption), the **"second quantum revolution"** underway is poised to deliver *positive* societal and business impacts far beyond cryptography. A recent World Economic Forum report emphasizes that quantum breakthroughs can **"unlock unprecedented growth opportunities"** for businesses ready to seize them. By 2035, quantum innovations could generate **trillions of dollars in value** across sectors such as finance, chemicals, life sciences, and transportation – a transformative wave of change that forward-looking leaders cannot afford to ignore.

Quantum Computing as an Innovation Catalyst in Industry

Quantum computing's unparalleled processing power will enable solutions to problems that are intractable for classical computers. This capability is set to **redefine entire industries** by enabling dramatically faster optimization, simulation, and analysis:

- **Energy and Climate:** Quantum computers can optimize complex systems like electrical grids and renewable energy integration, improving efficiency and sustainability. For example, quantum optimization could help balance power loads or improve battery efficiency for energy storage. In climate science, quantum simulations allow researchers to model vast, intricate systems (weather patterns, ocean currents, etc.) with unprecedented accuracy. These advances could enhance climate predictions and accelerate development of carbon capture and other clean technologies, offering powerful new tools in the fight against climate change.

- **Healthcare and Life Sciences:** In pharmaceuticals, quantum computing will transform drug discovery by enabling highly precise molecular simulations. Rather than relying on trial-and-error lab work, researchers can use quantum computers to model complex biochemical interactions and identify promising drug candidates much faster. This could **accelerate early-stage drug development** and reduce R&D costs. In healthcare, quantum algorithms and sensing devices also promise more accurate diagnostics – from modeling protein folding for disease research to pinpointing anomalies in medical imaging. Quantum sensors, for instance, may enable **earlier disease detection** through ultra-precise MRI and imaging techniques. Such capabilities would improve preventative care and treatment outcomes.

- **Financial Services and Economy:** The finance industry is anticipated to be one of the biggest beneficiaries of quantum computing. By 2035, use cases in finance (e.g. risk modeling, portfolio optimization, pricing complex derivatives) could create an estimated **$622 billion in value**. Quantum computers excel at solving optimization problems and running large-scale simulations – tasks ubiquitous in finance. They can analyze vast portfolios or market scenarios in parallel, finding optimal asset allocations or detecting subtle risk patterns that classical computers miss. Early experiments already show promise: institutions are exploring quantum algorithms for **faster derivatives pricing and more accurate portfolio risk assessments**. In short, quantum computing offers financial firms a competitive edge in decision-making, potentially **improving returns and risk management** beyond what was previously possible.

- **Manufacturing and Materials Science:** Quantum technologies will redefine how we design and produce the physical goods around us. A powerful near-term application is in materials science and chemistry – using quantum computers to **discover new materials and chemicals** with tailored properties. Quantum simulations can model atomic interactions in complex materials or chemical reactions far more accurately, accelerating the invention of advanced materials (for example, superconductors, high-strength polymers, novel catalysts) that could *dramatically* improve products and industrial processes. In the chemicals sector, this means optimizing reactions for higher yields or designing catalysts for greener manufacturing processes. Quantum-driven materials discovery is not science fiction: it could lead to breakthroughs in batteries, electronics, and industrial materials that unlock entirely new markets. As the World Economic Forum notes, each new material discovered via quantum computing has potential to **"spark growth in [its] industry, reshape supply chains and enable previously unattainable innovations."**

- **Transportation and Mobility:** The automotive and aerospace sectors are already tapping quantum tech to gain an edge. Quantum optimization algorithms can significantly improve complex logistics and design problems in transportation. For instance, Volkswagen has **used a quantum annealer to optimize bus routes**, minimizing travel time and fuel consumption by reducing traffic congestion. Likewise, quantum algorithms are being tested for dynamic traffic light optimization and fleet routing, which could ease urban traffic and save millions in fuel costs. In automotive R&D, major carmakers are leveraging quantum chemistry simulations to develop **next-generation electric vehicle batteries** with higher capacity and faster charging. Companies like BMW, Mercedes, and Hyundai have partnered with quantum computing firms to model battery materials and catalytic reactions at the quantum level, seeking breakthroughs in energy storage technology. Even manufacturing processes can benefit: BMW demonstrated quantum optimization for robot path planning on factory floors, hinting at more efficient production lines. In aerospace, quantum computing could help optimize aerodynamics and launch trajectories, or aid in designing lighter, stronger materials for aircraft. **Quantum sensing** is another game-changer for mobility – ultra-sensitive quantum GPS and magnetometers could enhance navigation and even enable navigation without satellite GPS by detecting minute gravitational variations. More precise sensors (e.g.

quantum lidar or accelerometers) promise improved autonomous vehicle performance and safety. In sum, quantum tech stands to make transportation systems smarter, greener, and more efficient from design through delivery.

Quantum Communications and Ultra-Secure Networks

Beyond computing, **quantum communication technologies** are emerging as a distinct sector that can revolutionize how we transmit information. Quantum communication promises **ultra-secure channels that are immune to hacking**, an advantage of immense value for protecting sensitive data and ensuring privacy. Unlike classical encryption (vulnerable to future quantum decryption), quantum communication methods such as Quantum Key Distribution (QKD) use the laws of physics to guarantee that any eavesdropping is immediately detected. In the coming years, we will likely see dedicated quantum-secure networks for governments, banks, and other organizations where secure communication is paramount – turning quantum-proof security into a competitive differentiator.

Looking further ahead, these advances pave the way for a full **quantum internet**. In a quantum internet, information (quantum states) would be transmitted between quantum devices, enabling capabilities far beyond secure messaging. For example, distributed quantum networks could allow **remote quantum computing**, where quantum processors in different locations work together on problems – effectively sharing quantum processing power over the network. Other envisioned applications include ultra-secure voting systems and quantum-enhanced sensor networks that leverage entanglement for coordinated measurements. Pilot projects are already demonstrating small-scale quantum networks in places like the Netherlands, China, and the US. In time, a mature quantum internet could underpin **entirely new services and industries** built on secure, instantaneous quantum information exchange – an infrastructure leap as significant as the original internet, but for quantum data.

Synergy of Quantum Technology and AI

A particularly potent vision for the future is the convergence of **quantum computing with artificial intelligence**. Quantum computing's ability to analyze enormous datasets and complex probability spaces could turbocharge AI and machine learning, enabling models of unprecedented sophistication. Researchers foresee quantum computers dramatically reducing training times for AI models by evaluating many scenarios in parallel. Hard optimization tasks in machine learning (like tuning millions of model parameters) might be solved more efficiently with quantum algorithms, leading to more powerful predictive analytics and AI capabilities. This *Quantum-AI* synergy cuts both ways: AI techniques are also being used to improve quantum technologies (for instance, optimizing error correction and qubit control). As one editor in the field put it, the blending of AI's adaptability with quantum's computational power **"offers a glimpse into a future where computational possibilities expand exponentially."** Entirely new industries could emerge from this fusion – for example, using quantum-augmented AI to discover drugs, materials, or designs that were completely out of reach before. From accelerating scientific research to enabling intelligent systems that can handle far more complexity, the quantum+AI combination will be a cornerstone of technological advantage in the coming decades.

Building the Quantum Advantage Beyond Security

In this forward-looking landscape, quantum technology is not just about countering threats – it's about **seizing opportunity**. Early adopters of quantum computing, sensing, and communication stand to gain significant competitive advantage by innovating new products and optimizing operations ahead of their peers. Governments and businesses worldwide are investing heavily now to cultivate quantum expertise, recognizing that leadership in quantum capabilities will translate to economic and strategic leadership tomorrow. The disruptive potential of quantum tech will not only *redefine* existing industries but also **create entirely new ones**, from quantum-enhanced health services to next-gen communications providers. In short, quantum technologies are set to become a catalyst for unprecedented economic and societal change – **not just shaping the future, but actively building it**. By looking beyond today's security concerns and embracing the broader quantum opportunity, organizations can position themselves at the forefront of the coming quantum-powered economy and unlock new advantages that once seemed beyond reach.

CLOSING MANIFESTO
BE AHEAD. BE SAFE. WIN.

In the quantum era, **leadership will make all the difference**. Quantum computing is no longer a sci-fi experiment or an IT puzzle for engineers – it's a strategic inflection point for businesses. This is *not* an IT issue; it's a leadership issue. The choices made in the boardroom today will determine who thrives in a world where quantum technology reshapes competition and security. **Inaction is a choice – and a risky one.** The window for leaders to seize the quantum advantage is open now, but it won't be open forever. Waiting for "certainty" or maturity in quantum tech is a **strategic error**. The message is clear: **don't wait for quantum to happen to you – lead the charge, starting now**.

Be Ahead — Anticipate, Experiment, Lead

Be Ahead of the curve. Great leaders anticipate disruption before it hits. Quantum computing's trajectory may be unpredictable, but its impact on business and security is inevitable. The executives who **anticipate** its effects on their industry and **experiment** early will be the ones out front. Some organizations are already investing tens of millions in quantum initiatives, refusing to sit idle. They're building quantum-readiness into their strategy, running pilots and forging partnerships to explore use cases. They know that **the first-mover advantage is real** – those who learn and adapt faster will leapfrog competitors stuck in "wait and see" mode. As one expert warned, *"waiting for quantum to be ready is a strategic error"*. Instead, prepare your people and your processes now. Foster a culture of curiosity and innovation around quantum. Encourage your teams to **experiment** with quantum solutions (from optimization to AI acceleration) and develop internal expertise. Set the tone that your company will be a **leader** in this emerging revolution, not a follower.

Staying ahead means **looking over the horizon**. Monitor breakthroughs and timeline signals closely – but don't rely on crystal-ball predictions. Plan for multiple scenarios (the breakthrough could come in 5 years or 15) and **stay agile**. Treat quantum as the next Internet or AI moment: a catalyst for those bold enough to embrace it early. The goal isn't to bet the company on unknown tech; it's to position your organization to capitalize when the quantum moment arrives. Leaders who act with urgency and foresight will shape the future, while laggards scramble to catch up. In short, **be the executive who sees the quantum wave coming and surfs ahead of it**. Your reward for being ahead is not just avoiding disruption – it's *owning* the opportunities that others will be too late to grasp.

Be Safe — Protect Trust, Harden Systems, Reduce Risk

Be Safe to safeguard what is non-negotiable: trust. Every executive knows that trust – with customers, partners, investors – is hard earned and easily lost. Quantum technology threatens to upend the cryptography and security foundations that underpin our digital trust. This isn't about sowing fear, but about taking *ownership* of the solution. The worst-case scenario is clear: today's encrypted secrets could be exposed tomorrow by a powerful quantum computer. In fact, adversaries are **already harvesting encrypted data today**, stockpiling secrets in hopes of decrypting them when quantum capabilities arrive. This "harvest now, decrypt later" tactic is *actively happening* and poses an "immediate and existential risk" to data privacy and security. We must respond with clarity and momentum, not panic. That means **protecting trust by acting early** to upgrade our defenses.

Start with your cryptography – it's the invisible bedrock of every digital transaction and communication. Quantum-safe encryption isn't a future nice-to-have; it's a present necessity. Standards for post-quantum cryptography (PQC) are emerging, and leading organizations are already beginning to weave them into their systems. Follow suit: **harden your systems** now by inventorying all the critical data, algorithms, and encryption in use. Develop a roadmap to transition to quantum-resistant cryptography in a phased, controlled manner. Remember that this transition is like **rebuilding the foundation of a skyscraper while people are still inside** – it must be planned carefully across the enterprise. That's why it demands *cross-functional* *coordination*: your cybersecurity teams, IT architects, compliance officers, vendors, and business unit leaders all need to be involved. A quantum-safe future can't be achieved in silos.

Most importantly, make "Be Safe" a core part of your company's ethos. Communicate to your stakeholders, employees and customers alike, that you are protecting their futures. By proactively reducing risk, you are preserving the trust they place in your organization. This is **risk management as a strategic leadership move**. It's not about fearing the quantum threat; it's about **neutralizing** it before it materializes. The companies that protect their data and systems now will not only avoid future crises, they'll also earn a reputation for reliability and foresight. In the quantum era, trust will be a competitive differentiator. Ensure that your organization remains worthy of that trust by being safe *today* so you can be secure *tomorrow*.

Win — Create Value, Differentiate, Transform

Ultimately, we strive to **Win**. "Winning" in the quantum era means creating new value, differentiating your business, and transforming your industry. Quantum advantage isn't just about defense against threats – it's about playing **offense** and seizing the immense opportunities this technology will unlock. The same quantum power that can threaten security can also solve problems once thought unsolvable. Leaders with vision are already asking: *How can quantum computing help us outperform our competition and serve our customers in revolutionary ways?*

The possibilities are truly game-changing. Quantum algorithms promise **exponential** leaps in computing capability. They could optimize complex supply chains and financial portfolios in seconds, model new drug molecules or materials with unprecedented precision, and crack optimization and AI challenges that stymie classical computers. Entire sectors stand to be reshaped. For example, in finance quantum computing may refine risk analysis and high-frequency trading; in pharmaceuticals it may accelerate drug discovery; in logistics it might route fleets in real time with optimal efficiency. These aren't distant fantasies – early prototypes have already demonstrated quantum's potential to **create value** in each of these arenas. As an executive, you should be continually scouting where quantum can open a door that was previously locked. **Differentiate** by aligning quantum innovation with your company's strategic strengths: if you're in logistics, pursue quantum optimization; if in pharma, invest in quantum chemistry research, and so on. Even a modest pilot project today can put you on the map as a forward-thinking leader in your field.

To win also means to **transform**. Just as the internet and AI forced businesses to reinvent themselves, quantum will reward those who rethink old problems and business models through a quantum lens. Position your organization not only to adopt new quantum solutions, but to *drive* quantum innovation in your market. This might mean partnering with a quantum startup, funding research, or building a small in-house quantum team to keep you at the cutting edge. Such moves signal to the market (and to your shareholders) that you intend to **shape the future** rather than react to it. They also ensure you won't be caught flat-footed when quantum breakthroughs accelerate. Remember, the goal isn't technology for technology's sake: it's to achieve meaningful business outcomes: better products, smarter services, more efficient operations, and delighted customers. Victory in the quantum age will go to those who combine technological prowess with strategic purpose.

Winning with integrity is key. We lead in quantum not to chase hype, but to deliver real, positive impact. With great power comes great responsibility – and quantum computing will be immense power. Use it wisely, guided by ethics and a commitment to improving security and society. If you **move with integrity**, you turn quantum advantage into lasting advantage, avoiding the pitfalls of misusing or overhyping the tech. In doing so, you differentiate your organization as one that not only *wins*, but wins the **right way** – with trust and vision at the core.

Lead from the Top — It's Your Mandate

None of the above happens by accident. It happens because **you**, as a leader, demand it and drive it. Quantum advantage requires boardroom ownership, executive accountability, and cross-functional orchestration. This is the moment for **boardroom accountability, cross-functional planning, and early motion**. Ensure your organization treats quantum not as a side project or a curious R&D experiment, but as a strategic priority woven into your overall vision. The board should be asking tough questions and tracking progress on quantum readiness, just as it would for any major business risk or opportunity. Rally your C-suite and top talent from every relevant function – strategy, IT, security, operations, finance – to develop a unified quantum action plan. Break down the silos: make it clear that quantum preparedness and innovation are company-wide efforts, not just the realm of the CIO or some specialist team. When everyone sees the **shared mission**, momentum follows.

Inaction is not an option. By now it should be evident that doing nothing is **the most dangerous move of all.** Every month of delay is a month adversaries and competitors edge ahead – and a month of potential vulnerabilities remaining unaddressed. Leadership means making the first move, even amid uncertainty. By taking *early action* – launching that pilot, initiating that cryptography overhaul, educating your teams – you create organizational muscle memory for agility and change. You send a signal to your stakeholders that you're not waiting to be disrupted or attacked – you're taking control of your destiny. This bias for action must start at the very top. If not you, who? If not now, when?

The good news: **the advantage is still yours to seize**. We are at the dawn of the quantum age; most companies have barely begun. You have a rare opportunity right now to leap forward while others hit snooze. But this window will close. Soon, quantum risk management will be mandated, and quantum innovation will be table stakes. Latecomers will find themselves scrambling to retrofit security and play catch-up in the marketplace. **That fate is avoidable – if you lead decisively, now**. Foster a sense of urgency in your organization, but not a sense of fear. This is a race *you can win* by acting with clarity, speed, and integrity. It's about doing the right things *before* you're forced to, and positioning your enterprise to thrive when the quantum disruption fully materializes.

In closing, the mandate is simple and bold: **Be Ahead. Be Safe. Win.** These are more than slogans – they are principles for action. Anticipate the future and shape it (don't wait for it). Secure your digital foundation to uphold trust (don't gamble it). And create new value boldly (don't just defend, *innovate*). The quantum era calls for leaders who are proactive, not reactive; visionary, not complacent. You, as an executive reader equipped with insight and urgency, are exactly the leader to answer that call. **The advantage will go to those who move with speed, integrity, and vision** – so set the pace, set the standard, and seize your quantum advantage.

Executive Signal: The quantum future rewards the bold. Lead with foresight and fortitude – make the first move, secure your foundations, and transform your business. **Be Ahead. Be Safe. Win.**

APPENDICES

APPENDIX 1:
GLOSSARY OF QUANTUM AND PQC TERMS

APPENDIX 2:
QUANTUM-SAFE READINESS CHECKLIST

APPENDIX 3:
GLOBAL MANDATES & STANDARDS TIMELINE

APPENDIX 4:
SECTOR USE CASE SNAPSHOTS

APPENDIX 1: GLOSSARY OF QUANTUM AND PQC TERMS

- **Post-Quantum Cryptography (PQC):** Also called *quantum-resistant cryptography*, this refers to new cryptographic algorithms designed to withstand attacks by both classical and quantum computers. These algorithms (e.g. lattice-based encryption and signature schemes) aim to replace today's RSA and ECC, so that even a powerful quantum computer cannot crack the encryption in feasible time. PQC is essential for protecting data against the future quantum threat.
- **Quantum Key Distribution (QKD):** A method of securely distributing encryption keys using quantum physics. QKD uses properties like the *no-cloning theorem* of quantum particles to let two parties share a secret key – any eavesdropping on the quantum channel can be detected. QKD can provide theoretically secure key exchange, though it has distance and infrastructure limitations.
- **Hybrid Cryptography:** In the quantum-safe transition, *hybrid* schemes combine traditional public-key algorithms with PQC algorithms in one protocol. For example, a TLS connection might use both an RSA/ECC key exchange and a post-quantum key exchange together. This offers **transitional resilience** – even if one method is broken, the other remains, ensuring the security of the channel. Hybrid approaches are a pragmatic step to gradually integrate PQC while retaining classical security.
- **Quantum Supremacy (Quantum Advantage):** A milestone in computing where a quantum computer demonstrably outperforms the most powerful classical computers on a specific task. In practice it means solving a problem that no conventional computer could solve in any reasonable timeframe. *Quantum supremacy* was first claimed in 2019 when Google's quantum processor solved a contrived problem exponentially faster than a supercomputer. It showcases the *potential* of quantum computing, though these early demonstrations are not yet practical business use cases. The broader term *quantum advantage* is often used when quantum computing provides a significant performance benefit on a real-world application, even if classical computing isn't completely left

in the dust.

- **Crypto-Agility:** The ability of an organization's systems and processes to **swiftly swap out cryptographic algorithms** with minimal disruption. High crypto-agility means you can upgrade or replace encryption (for example, switching from RSA to a PQC algorithm) in your applications, devices, and communications quickly as threats evolve. This capability is vital for a smooth transition to post-quantum encryption – it ensures that as soon as new secure algorithms or standards emerge, the organization can adopt them without lengthy redesigns. Crypto-agility is now considered a required best practice to stay ahead of both quantum threats and any future advances that might weaken current crypto.

APPENDIX 2: QUANTUM-SAFE READINESS CHECKLIST

Use this checklist to ensure your organization is prepared to migrate to quantum-safe cryptography. The checklist spans key domains – from taking stock of your cryptographic assets to mobilizing your teams – as identified by industry experts. Each item represents an action area for executives and strategy leaders:

- **Leadership & Governance:** Appoint a clear **owner for quantum risk** (e.g. a *cryptographic migration lead*) within your organization. Establish executive oversight through a steering committee or include quantum readiness in an existing risk committee. Develop a high-level **quantum security roadmap** with board support, setting target dates aligned with regulatory deadlines and the anticipated arrival of quantum threats. Ensure quantum risk is part of your enterprise risk management and regularly reported.
- **Cryptography Inventory:** Conduct a thorough **inventory of all cryptography use** in your systems and products. Identify where and how data is encrypted or digitally signed – including software applications, network protocols (VPNs, TLS), databases, IoT devices, etc. For each, document the algorithms (e.g. RSA-2048, ECC P-256, AES-128) and key lengths in use. Prioritize "High-Value Assets" and sensitive data that must remain secure for many years. *(This "crypto-BOM" – bill of materials – will guide your transition efforts.)*
- **Risk Assessment & Prioritization:** Assess which systems and data are most **vulnerable to quantum attack** and what the impact would be if encryption failed. Prioritize systems that protect long-lived secrets or critical data (for example, medical records, intellectual property, financial transactions) because of "**harvest now, decrypt later**" risks. Develop a **phased migration plan**: for instance, plan to secure critical customer data and inter-bank communications before less sensitive systems. Use the inventory to create a timeline – e.g. "quantum-proof" highest-risk applications by 2030, secondary systems by 2035 (in line with global guidance).
- **Cryptographic Agility:** Invest in making your

infrastructure **crypto-agile** now. This means updating software, hardware, and vendors so that algorithms can be changed with minimal disruption. For internal applications, refactor or layer abstraction to avoid hard-coding specific ciphers. Introduce support for new PQC algorithms (many libraries already offer them in test form). Implement **hybrid encryption modes** (e.g. TLS 1.3 with a hybrid key exchange) where available to trial PQC alongside classical crypto. Ensure your upgrade cycles (for hardware security modules, VPN appliances, etc.) include models or firmware that support PQC. The goal is to be ready to "drop in" quantum-safe algorithms as they become standardized.

- **Pilot Projects & Testing:** Plan and execute **PQC pilot projects**. For example, select a non-production system or internal tool to implement a post-quantum encryption algorithm (like NIST's CRYSTALS-Kyber for key exchange) and observe the performance and integration challenges. Run interoperability tests: can your systems communicate using PQC with partners or customers? Pilot a **hybrid solution** (combine classical and PQC) to familiarize your IT staff with the new tech and to reveal any issues early. These pilots will build organizational know-how and confidence, letting you refine your broader roll-out plan.
- **Vendor Engagement: Press your vendors and partners** to act. Ask critical suppliers (cloud providers, VPN/firewall vendors, core banking software providers, etc.) about their quantum-safe roadmap – **when will their products support PQC or hybrid modes?** Include PQC readiness as a criterion in new procurements and renewals. For current vendors, insist on timeline commitments or updates (many governments and large companies are doing this now). Collaborate in industry groups or standards bodies (if relevant) to encourage a coordinated transition. By signaling demand for quantum-safe solutions, you not only protect your supply chain but also help set industry expectations.
- **Training & Team Formation:** Build an internal **quantum-ready team** or working group that brings together IT security, cryptography experts, application owners, and risk managers. Conduct awareness sessions for developers and architects on the coming quantum threat and post-quantum solutions. Offer targeted training on implementing new cryptographic libraries or protocols. Identify external partners or consultants who can assist,

if you lack in-house cryptographic expertise. Ensuring your people are educated will smooth the migration – the human element is as important as the tech.

- **Policy & Compliance Update:** Update your security policies and IT guidelines to incorporate **quantum-safe practices**. For example, add requirements for crypto-agility in solution designs, mandate that new systems use approved PQC algorithms (once standards are final), and include quantum risks in data protection policies. Monitor emerging **standards and mandates** (e.g. NIST standards, ISO standards, government regulations) and ensure your compliance team is tracking these. Align your roadmap with any **regulatory deadlines** (see Appendix 3) to avoid last-minute scramble.

- **Ongoing Monitoring:** Treat quantum readiness as an ongoing program. Keep watching **technology progress** – both the advance of quantum computing capabilities and the evolution of PQC algorithms. Regularly revisit your risk assessment: if breakthroughs occur (e.g. a new quantum algorithm that speeds up attacks), adjust your timelines accordingly. Similarly, monitor for **standards updates** from NIST, ISO, IETF, etc., and for any vulnerabilities discovered in the new PQC algorithms (the field is new, so this is possible). Have a process for updating your cryptographic toolkit if needed (e.g. replacing one PQC algorithm with another if weaknesses are found).

Checklist Tip: Consider integrating this checklist into your project management tracker. For each item, assign an owner and a target date. For example, *"Complete crypto inventory – Owner: CIO office – Due: Q2 2024."* This makes the abstract threat actionable. By covering governance, technology, people, and partners, you ensure no facet of the quantum-safe transition is overlooked.

APPENDIX 3: GLOBAL MANDATES & STANDARDS TIMELINE

Quantum-safe cryptography is now on the agenda worldwide. Below is a timeline of key global mandates, standards, and regulations pushing the transition to PQC. Executives should be aware of these milestones, as they often set **deadlines for action** or signal when quantum-safe solutions will become the norm.

- **May 4, 2022 – United States:** The White House issues **NSM-10 (National Security Memorandum 10)**, *"Promoting U.S. Leadership in Quantum Computing While Mitigating Risks to Vulnerable Cryptography."* This directive mandates that federal agencies begin transitioning to post-quantum cryptography and aims for most quantum-vulnerable systems to be mitigated by **2035**. It signals top-level U.S. commitment: agencies must *"prepare now"* for a post-quantum world.
- **July 5, 2022 – United States: NIST announces** the first group of winners in its PQC competition, selecting algorithms like **CRYSTALS-Kyber** (encryption key exchange) and **CRYSTALS-Dilithium** (digital signatures) for standardization. This effectively identifies the **future standards** to replace RSA and ECC, giving industry a heads-up on which new algorithms to implement.
- **September 2022 – United States (NSA):** The National Security Agency releases **CNSA 2.0 (Commercial National Security Algorithm Suite 2.0)** – new cryptographic standards for U.S. National Security Systems that include quantum-resistant algorithms. It sets a timeline where, for example, **by January 1, 2027, all new national security systems must use CNSA 2.0 compliant (quantum-safe) crypto**, and by 2033 existing systems must transition. This is one of the first detailed "must-use-PQC" mandates, though focused on defense and intelligence systems.
- **November 18, 2022 – United States (OMB):** The U.S. Office of Management and Budget issues **Memo M-23-02** "Migrating to Post-Quantum Cryptography." It requires every federal agency to **inventory its cryptographic systems by May 4, 2023** and annually thereafter through 2035. Agencies must identify where they use cryptography and report on progress – essentially kick-

starting migration efforts. This memo also required agencies to name a crypto-transition lead within 30 days. (OMB will later issue more guidance once standards are finalized.)

- **December 2022 – United States:** The **Quantum Cybersecurity Preparedness Act** is signed into law. It reinforces OMB's efforts by requiring OMB to **provide Congress with a roadmap and guidance for PQC adoption within one year of NIST finalizing standards**. (Since NIST finalized standards in Aug 2024, this means OMB guidance by late 2025.) This Act underlines bipartisan concern about quantum threats and ensures long-term oversight of federal migration.
- **November 2022 – European Union:** The EU formally adopts the **NIS2 Directive**, a major cybersecurity directive for critical industries. NIS2 doesn't explicitly mandate post-quantum crypto yet, but it requires organizations in critical sectors to practice "state of the art" security and cryptographic risk management. It came into effect in 2023, and EU Member States must transpose it into national laws by **October 17, 2024**. This effectively means that by late 2024, banks, energy companies, healthcare providers and others under NIS2 will need to consider emerging risks like quantum and have plans (crypto-agility could be seen as part of "state of the art" practices).
- **March 2023 – Australia:** The Australian Signals Directorate (ASD) and Cybersecurity Centre publish **"Planning for Post-Quantum Cryptography"** guidance. Notably, Australia set an **aggressive 2030 target** – stating that **by 2030, suppliers of High Assurance cryptographic equipment must use quantum-resistant solutions**. In other words, Australia signaled a cutoff: after 2030, government will not accept products with legacy encryption for high security applications. This ups the pressure on vendors and government agencies alike in the next few years.
- **March 2023 – Japan:** Japan's CRYPTREC (cryptography research center) issues **"Cryptographic Technology Guidelines: Post-Quantum Cryptography."** This 2023 document provides a roadmap for Japanese government and industry to begin using PQC algorithms. Japan has been running its own PQC trials and will align with many NIST-chosen algorithms. Expect Japanese government procurement standards

to start including PQC in the mid-2020s.

- **August 2024 – United States: NIST publishes the first set of official PQC standards.** In August 2024, NIST released FIPS 203, 204, and 205 – standards for post-quantum key establishment (encryption) and digital signatures. NIST urged administrators to *"begin transitioning to the new standards as soon as possible"*. This is a landmark: the "starting gun" for broad adoption. Many regulations (like the U.S. Act above) tied their next steps to this event. With standards in hand, vendors can finalize products, and agencies can proceed with concrete implementations knowing they're using approved algorithms.
- **February 20, 2024 – Singapore:** The Monetary Authority of Singapore (MAS) issues **Advisory Circular MAS/TCRS/2024/01** to financial institutions on "Addressing the Cybersecurity Risks Associated with Quantum." MAS urges banks and insurers to **maintain a cryptographic inventory and achieve crypto-agility** in preparation for quantum threats. It recommends that organizations start raising awareness of quantum risks, prioritize which crypto systems to upgrade, and explore solutions like PQC and QKD. While not a fixed deadline, this guidance effectively tells the finance sector in Singapore: *"start now on quantum-safe transition."*
- **April 11, 2024 – European Union:** The European Commission, with ENISA and the NIS Cooperation Group, publishes a **Recommendation and Roadmap for Transition to PQC**. This coordinated roadmap (finalized in a document by June 2025) calls for Member States to begin migrating critical infrastructure to post-quantum cryptography **by 2025**, and to have **priority use-cases secured by 2030**. It further urges that *"as many systems as possible"* be quantum-resistant by **2035**. In effect, the EU set 2030 as the first checkpoint (for high-risk systems like government, finance, telecom) and 2035 as an end goal for broad adoption.
- **March 20, 2025 – United Kingdom:** The UK's National Cyber Security Centre (NCSC) releases new guidance outlining **timelines for migration to PQC**. It breaks the transition into three phases: (1) **Discovery by 2028** – organizations should complete crypto discovery/inventory and start testing PQC by then; (2) **Prioritized Upgrades by 2031** – migrate the most important systems (those critical to security or protecting long-lived sensitive data) to PQC

by 2031; (3) **Complete Migration by 2035** – transition all remaining systems and products to quantum-safe cryptography by 2035. This timeline aligns with NIST's deprecation goals and gives UK organizations a clear, staged plan with target dates.

- **Late 2025 – United States:** Under an updated cybersecurity executive order (EO 14144, 2024) and the Quantum Cybersecurity Act, U.S. agencies will receive further instructions. Notably, by **December 1, 2025**, NSA and CISA must publish a list of approved **"quantum-safe" product categories** (e.g. identifying what types of security products should be PQC-capable). Additionally, the U.S. government set a requirement for all federal agencies to adopt **TLS 1.3 (or a later protocol) by January 1, 2030** – this ties into quantum safety because TLS 1.3 can more easily implement hybrid/PQC cipher suites. The government is essentially saying: upgrade your networks now, so they're ready for post-quantum crypto well before 2035.
- **Beyond 2025 – Ongoing:** Many other countries are following suit. **Canada** in 2025 published a government migration roadmap similar to the UK's. **Germany, France, Netherlands** and others have issued national PQC recommendations (often aligning with NIST choices, but sometimes including additional algorithms). Standards bodies like **ISO** and the **IETF** are standardizing PQC for protocols (for example, the IETF is working on hybrid TLS extensions). The general consensus timeline, often called **"Q-Day" or "Y2Q"**, is roughly **2030-2035** for when quantum computers could pose a real threat – hence, governments worldwide are pushing organizations to be quantum-safe by around 2030-2035 at the latest. Executives should keep an eye on local regulations and international standards through the rest of the 2020s.

APPENDIX 4: SECTOR USE CASE SNAPSHOTS

Every industry will experience the quantum revolution a bit differently. Below are brief snapshots for four sectors – **Finance, Pharmaceuticals, Logistics, and Cloud Computing** – highlighting the key opportunities quantum tech presents, the threats to be mitigated, and a rough timeline of developments to watch. The focus is on **practical, strategic insights** for leaders in each sector.

Finance (Banking & Financial Services)

- *Opportunities:* Financial institutions could harness quantum computing for advanced **portfolio optimization, risk modeling, and fraud detection**. Problems that are computationally intense today – like analyzing countless investment strategies or optimizing trading algorithms – may be solved faster with quantum algorithms. For example, quantum processors excel at certain optimization and sampling tasks, which can improve **market trend predictions and option pricing** in ways classical computing can't easily match. Some banks are already experimenting with quantum-based Monte Carlo simulations for pricing derivatives and machine learning for detecting fraudulent transactions. Early advantages might also come from quantum-enhanced secure communications; e.g. **quantum key distribution between banking data centers** to ensure ultra-secure transactions.

- *Threats:* The most immediate threat is to the encryption that underpins financial systems. **Quantum attacks could break RSA-based security**, imperiling everything from secure banking websites to inter-bank fund transfer networks. A successful quantum attack in the future could mean an attacker decrypting customer account data, payment transactions, or even blockchain-based assets (cryptocurrencies). The industry is also vulnerable to "**harvest now, decrypt later**" espionage – where adversaries steal encrypted financial data now (for example, intercepting transaction records or SWIFT messages) with the intent to decrypt it once a quantum computer is available. Additionally, the integrity of algorithms used in high-frequency trading or automated decision-making could be at risk if quantum technology gives some players a computational edge (raising future regulatory and fairness questions). In short, quantum computing could both undermine the security **and** upset the competitive balance of financial markets if some firms gain access earlier.

- *Timeline:* Financial regulators are taking notice. Expect **regulatory guidance by late 2020s** requiring banks to inventory and upgrade their cryptography (the U.S., UK, and Singapore authorities already have advisories in place). By **2030**, many large financial institutions aim to have critical systems quantum-resistant, in line with global mandates (see Appendix 3). On the opportunity side, **5–10 year horizon** is often cited for practical quantum advantage in finance. Already, **by 2025**, we see prototypes – e.g. Mastercard and JPMorgan doing pilot projects with quantum providers. Between **2025 and 2030**, expect incremental gains: particular use-cases like fraud detection may see quantum-inspired algorithms improving results even before true quantum computers are fully ready. **By the early 2030s**, if hardware progress continues, banks might use quantum computers (likely through cloud access) for mainstream risk analytics, achieving faster or more accurate results than classical HPC can deliver. Executives in finance should watch central bank initiatives (like the Bank of England and ECB experiments) and be prepared for regulators to set hard requirements around 2030 for quantum-safe encryption of financial data.

Pharmaceutical & Healthcare

- *Opportunities:* Quantum computing holds transformative potential in pharma. The most headline-grabbing possibility is **quantum-accelerated drug discovery and molecular modeling**. Quantum computers can, in theory, simulate complex molecular interactions at a level of detail impossible for classical computers, which could lead to discovering new drug candidates and optimizing compounds much faster. For pharma companies, this means solving problems like protein folding, or screening billions of molecules against a target, with unprecedented speed – potentially cutting R&D timelines significantly. Beyond drug discovery, quantum algorithms could improve **genomic analysis**, help design personalized medicine by analyzing vast genetic datasets, and optimize clinical trial logistics (like more efficient patient stratification). Even the healthcare sector could benefit through quantum-improved machine learning for diagnostics (e.g. imaging analysis). In short, quantum tech offers a chance for a *step-change* in innovation and efficiency for life sciences.

- *Threats:* The pharma and health sector also rely on data confidentiality and integrity. **Intellectual property (IP)** such as drug formulas, research data, and patient health records must remain secure for decades – all of which is currently protected by encryption that will become vulnerable. An pharma company's valuable encrypted research data could be stolen now and decrypted in a decade by a competitor or malicious actor, eroding competitive advantage. Also, many medical devices and healthcare IT systems use public-key cryptography for updates and authentication; these could be compromised by quantum attackers, leading to counterfeit medical devices or data breaches. There's also a **nation-state threat**: a government with quantum capability might target pharmaceutical IP (vaccines, for example) for economic or strategic gain. Moreover, *if* quantum computing helps competitors discover drugs faster, companies that lag behind could see their products become obsolete quicker – a competitive threat. Lastly, any *perception* that health data isn't secure (due to quantum cracks) could undermine patient trust, so the sector must be proactive.

- *Timeline:* The timeline for quantum opportunity in pharma is often cited as **mid to late 2020s for initial breakthroughs**. Recent roadmaps suggest that industries like pharma might see **practical quantum advantage sooner** than others – possibly within **5–10 years** for certain chemistry problems. For example, by around 2030, we might witness a quantum computer correctly modeling a complex biochemical reaction that classical supercomputers cannot. Pharma giants and biotech startups are already partnering with quantum hardware companies (e.g., for simulation of protein-ligand interactions). On the threat side, the need to secure data is *immediate*: given the long sensitivity of research data, firms are advised to start deploying quantum-safe encryption in the next few years (2025–2028 for critical data repositories) to ensure data collected now remains safe through the 2030s. We will likely see **health sector regulators issuing PQC guidelines by late**

2020s, possibly sooner for areas like medical devices (imagine an FDA or EMA requirement that devices use quantum-safe firmware signing by a certain date). By **2035**, both the quantum computing-enhanced drug discovery and the full migration to quantum-safe crypto in healthcare should be well underway, if not earlier, given the value of the intellectual property at stake.

Logistics & Supply Chain

- *Opportunities:* Logistics is an industry of optimization, and quantum computing is poised to be an optimization powerhouse. Use-cases include **route optimization** (finding the absolutely best shipping routes and delivery schedules among millions of possibilities), **supply chain design** (optimally placing warehouses and allocating inventory), and complex scheduling (like port loading sequences or aircraft scheduling). Quantum algorithms (especially quantum annealing or variational algorithms) are naturally suited to these *combinatorial optimization* problems. A quantum computer could potentially compute an optimal delivery network configuration that a classical system can only approximate. Companies like DHL and FedEx have already dabbled in quantum-inspired algorithms for route planning. Another opportunity is **quantum-enhanced demand forecasting** – using quantum machine learning on large datasets to better predict demand spikes and adjust supply chain parameters in real time. Over a long global supply chain, even a few percentage points of efficiency gained can save millions of dollars, so the incentive is high. Additionally, quantum sensors (a different branch of quantum tech) could improve logistics by providing ultra-precise navigation (quantum GPS alternatives) or detecting environmental conditions during transport with greater accuracy – though those are more on the hardware side of quantum technology.

- *Threats:* The logistics sector increasingly depends on digital systems and global data exchange (think of the logistics APIs and platforms connecting manufacturers, shippers, ports, and retailers). These rely on cryptography for authentication and secure data sharing. **Quantum threats to logistics** include potential breaches of cargo tracking systems or tampering with shipping manifests if cryptographic signatures are broken. For instance, bills of lading or custom documents could be forged if digital signature algorithms (like ECDSA) are cracked – leading to fraud or theft in the supply chain. The sector is also part of critical infrastructure; a quantum-enabled attacker could disrupt logistics operations (e.g., take down encrypted communication between control centers and ships/planes). Furthermore, blockchain is making inroads in supply chain provenance tracking – and blockchains largely use classical crypto, which quantum could undermine (imagine falsifying the provenance of a product by breaking the chain's crypto). In summary, an insecure supply chain can be chaotic: goods could be stolen or rerouted and data could be manipulated, causing financial losses and safety issues. The **economic incentive for cybercriminals** (and state actors) to target global logistics means this sector must not be the slowest to upgrade its security.

- *Timeline:* Many logistics companies will likely follow the lead of broader industry and government requirements. There may not be sector-specific quantum mandates yet, but because logistics underpins other critical sectors (food, medical supplies, defense logistics), we anticipate **quantum-safe guidelines by late 2020s** for large logistics operators, possibly as part of critical infrastructure protection rules. By **2030**, expect that major shippers and port authorities will require quantum-safe encryption in their data interchanges – possibly driven by regulations in the EU (NIS2 covers transportation) or the US (Transportation Security Administration guidance, etc.). On the opportunity front, **quantum optimization trials are already happening**. As quantum hardware scales this decade, a practical benefit might appear in **route optimization or supply scheduling by around 2030** – perhaps a pilot where a quantum solver saves a global shipper significant fuel and time. Between 2030–2035, as quantum computers become more powerful, the logistics industry could see a gradual but steady adoption: first for strategic planning (network design, seasonal forecasting), and later for real-time dynamic routing. Executives should watch partnerships between tech providers and logistics giants (e.g., any announcements from UPS, Maersk, etc. on quantum experiments) as bellwethers. Also important: incorporate quantum-safe standards into long-term procurement now (e.g., if buying fleet communication systems in 2025, ensure they are crypto-agile or upgradable).

Cloud & Tech Providers

- *Opportunities:* The Cloud Computing sector is uniquely both an **enabler** of quantum technology and a **consumer** of quantum-safe solutions. On one hand, major cloud providers (such as IBM, Google, Microsoft, Amazon) are heavily investing in quantum computing R&D and even offering **cloud-based quantum computing services**. This means that enterprises can access prototype quantum computers via the cloud now – for instance, to experiment with quantum algorithms without owning any hardware. This "Quantum Computing as a Service" model will likely accelerate innovation in all other sectors by democratizing access to quantum resources. On the other hand, cloud providers can integrate **quantum-safe cryptography at massive scale**, upgrading the security of cloud storage, databases, and communications for all customers. In fact, cloud companies often lead the charge in implementing new security tech – we're already seeing cloud services support PQC algorithms in test mode. For example, some cloud platforms have enabled **hybrid TLS configurations** (mixing classical and PQC key exchange) for secure connections, anticipating future standards. Cloud data centers might also deploy quantum random number generators to strengthen cryptographic keys. In sum, the cloud sector's opportunity is twofold: profit from offering quantum-powered services (quantum computing, quantum randomness, etc.) and differentiate by providing **quantum-resistant security** to clients early. This can be a marketable feature, assuring customers that their data will remain safe into the quantum era.

- *Threats:* Cloud providers manage vast amounts of other companies' data, making them prime targets. If a cloud platform's encryption were broken, the breach could be catastrophic, affecting many businesses at once. **Cloud infrastructure heavily relies on public-key cryptography** for everything from user authentication, to securing data in transit (HTTPS, VPNs), to protecting data at rest. A future quantum adversary could theoretically decrypt intercepted cloud traffic, impersonate identity providers by breaking certificates, or access encrypted databases. The **zero trust models** that cloud vendors and enterprises are adopting would be undermined if the cryptographic trust anchors (like digital certificates and tokens) can't be trusted. Additionally, the cloud sector must worry about the **transition pain**: updating thousands of systems and services to new cryptographic standards is a huge task – any incompatibility could cause downtime or security gaps. There's also a competitive threat: if one cloud provider is perceived as lagging in quantum-safe security, customers (especially in regulated industries) might migrate to a competitor. Finally, consider the scenario where nation-state actors prioritize breaking cloud security (since "that's where the data is") – cloud companies might face sophisticated quantum-enabled attacks sooner than others.

- *Timeline:* The timeline here is aggressive. **Major cloud providers are already acting**. For instance, **Microsoft launched a Quantum Safe Program in 2025 aiming to upgrade its infrastructure and services to PQC by 2033** – intentionally ahead of many government deadlines. Google and AWS likewise have internal efforts to integrate PQC (some have participated in NIST's process and are updating open-source crypto libraries). We will see **client-facing features** sooner: by 2024–2025, expect cloud vendors to start announcing that connections to their services (VPNs, web endpoints) can optionally use post-quantum TLS, and that customer data encryption options include PQC algorithms. By **2028**, many cloud customers (especially government clients) will demand evidence of quantum-safe security in RFPs and contracts – pushing cloud providers to have completed a large part of the transition. On the quantum computing service side, the next few years (2025–2030) will see rapid improvements in available quantum hardware via the cloud (e.g., IBM's roadmap to thousands of qubits by 2025-2026). This means enterprises will start to get real value from cloud quantum services potentially by the late 2020s – for example, cloud-based quantum ML for big data analytics. **By 2030**, cloud platforms might offer turnkey quantum-safe cryptography options (flip a switch and all your storage buckets use PQC encryption, for instance) as well as advanced quantum computing resources. The cloud sector, due to its scale and central role, will likely be *ahead* of most others in migrating (aiming to finish by ~2030 for core systems, aligning with the stricter government mandates). For executives relying on cloud providers, the key is to **engage with your cloud vendors early** – inquire about their quantum-safe roadmap and use those features when available, as it may drastically simplify your own organization's quantum security journey.

In Summary: Each sector has unique drivers for quantum adoption and urgency for quantum defense, but a common theme emerges: **the time to prepare is now**. Whether it's to seize new computational advantages or to safeguard critical assets, leading organizations in finance, pharma, logistics, and tech are already incorporating quantum into their strategic plans. Watching the timelines (for technological milestones and regulatory requirements) will help you stay ahead, be safe, and ultimately win in the quantum era.

ABOUT THE AUTHOR

Mike Mikhail is a technology advisor and security expert with deep experience in emerging technology transitions.

Mike currently serves as a Delivery Architect in Cisco's Customer Experience (CX) organization, where he helps guide enterprises through cybersecurity challenges and digital transformations.

In **Quantum Advantage**, Mike distills complex quantum science into practical strategies for executives, emphasizing his motto "**Be Ahead. Be Safe. Win.**" to ensure organizations thrive securely in the quantum age.

www.ingramcontent.com/pod-product-compliance
Lightning Source LLC
Chambersburg PA
CBHW062028200326
41519CB00017B/4973